DECONVOLUTION

DECONVOLUTION

ANTON ZIOLKOWSKI

DELFT UNIVERSITY OF TECHNOLOGY

INTERNATIONAL HUMAN RESOURCES DEVELOPMENT CORPORATION
BOSTON

Library of Congress Cataloging in Publication Data

Ziolkowski, Anton, 1946–
  Deconvolution.

  Bibliography: p.
  Includes index.
  1. Seismic reflection method   Deconvolution.
I. Title.
TN269.Z57 1983   622'.159   83-12624
ISBN 0-934634-62-9

Printed in the United States of America

TO BILL LERWILL

CONTENTS

PREFACE

This book is about seismic deconvolution. It describes what is normally done to deconvolve seismic data and examines in detail whether the assumptions that are made can be justified in terms of the model.

Whenever deconvolution is applied, a convolutional model is invoked. It turns out that two mutually exclusive models are in use: a point-source model and a plane-wave model. The choice of model is governed not by the field data acquisition parameters, as one might expect, but by whether the source signature is known or not. If it is known, the point-source model is assumed; if it is unknown, the plane-wave model is assumed. Obviously the choice of convolutional model in itself is a problem.

I have often been told that the convolutional model is all wrong anyway, so one must accept that some of the processes we apply cannot be justified strictly in terms of this model. To this I reply: If the convolutional model is no good, what is the alternative model? When we manipulate data we must be able to justify to ourselves why it makes sense to do so. Very often the response to this is: Even if we cannot explain *why* something should work, when we can see that it *does* work, we should do it. Still suspicious, I come back with the argument that, simply because the data look "better" in some sense, it does not follow that we are any closer to the truth. At this point, the quality of the argument always deteriorates into a discussion about what is "the truth."

I have never been able to get any further in this argument than these few steps and it is a very frustrating business. This sense of frustration has been the principal motivation for the writing of this book, and I apologize if, here and there, my voice is raised a little too loudly in the presentation of my case.

I have tried to concentrate on the theory of what is normally done in seismic deconvolution, which is generally regarded as a data processing problem, rather than as an acquisition or interpretation problem. The processing geophysicist tries to obtain the maximum resolution of the primary reflections from the recorded seismic reflection data. To do this he or she is equipped with a number of computer programs that extract a wavelet from the data, calculate a filter from the wavelet, and apply the filter to the data to remove the wavelet. The final step is known as deconvolution. With a given data set, the result of deconvolution depends entirely on the method of estimation of the wavelet. Each method yields a different result. The problem for the processing geophysicist is to choose between the various deconvolution methods and to choose the parameters for the estimation of the wavelet with the chosen method.

This book investigates the difficulties in making these choices and shows that the problem, as posed, is insoluble. The fundamental difficulty is that processing geophysicists are not provided with enough information to solve the problem. We must also be given accurate information about the input source wavefield. In seismic reflection we have a great deal of control over what we do: we not only record the echoes, we also arm and fire a controlled seismic source. The wavefield generated by the seismic source must be measured. How we should measure this wavefield is not the subject of this book (although it does deserve a book to itself). The point is that, unless this source information is provided, the process of deconvolution has to rely on guesswork. There is a whole class of deconvolution methods that depend on guesses about the statistical properties of the earth impulse response in order to find out what the impulse response of the earth really is. Furthermore, within some of these methods, it is shown here that the guesses or assumptions are mutually inconsistent.

I never underestimate the success that good guesses have in pushing forward the science of geophysics; I simply argue that to make guesses instead of measurements the standard procedure may not necessarily lead to reliable answers.

This book, then, is more of an argument than a text. Where a digression does not help the argument, I have tried to avoid it. As a consequence, certain topics that one would perhaps expect to see covered in a textbook on deconvolution are not included here. This book should be treated as a supplement to the more optimistic books and papers on the subject, especially those that have helped to give the process of deconvolution an aura of magic.

Anton Ziolkowski

Delft, November 1983

Acknowledgments

My interest in deconvolution began at Cambridge in 1968, when my supervisor, Fred Gray, introduced me to a recently published book, by Enders Robinson, entitled *Statistical communication and detection with special reference to digital data processing of radar and seismic signals*. This book, which opened up a whole new world for me, was so lucid and such a joy to read that I eagerly read everything I could find by Enders Robinson. I soon found most of the papers by Robinson and Treitel, which were already classics. The papers were generally harder to read than Robinson's book, which I needed to go back to time and again to try to understand the papers. Without this book and without these papers, I would never have developed an interest in deconvolution.

It was so interesting that I was always trying to tell other people about it. The first who were kind enough to listen were Dan McKenzie, Andrew Stacey, Dave Gubbins, and Carol Williams. Later on, when I left Cambridge for the Massachusetts Institute of Technology, I found how shaky my understanding really was. It was at MIT that I learned a little more about the theory and something about the practice of manipulating time series. I was greatly helped in this by Dick Lacoss, Tom Landers, and especially Clint Frasier.

In 1973, shortly after I left MIT to go the London School of Economics, Paul Stoffa sent me his thesis on homomorphic deconvolution, which seemed to me to be a real breakthrough in the approach to the deconvolution problem. He shifted the minimum-phase assumption from the wavelet to the earth-impulse response and allowed the wavelet to be mixed phase; I thought this was much more sensible than the conventional approach.

Over the next few years, I benefited from many discussions about deconvolution with John Beavan, Elio Poggiagliolmi, Dave Brown, Paul Stoffa, Les Hatton, Charlie Hewlett, Greg Beresford-Smith, Neil Goulty, Chris Walker, Lloyd Peardon, and Derek March. During this time I became gradually more and more skeptical about the usual seismic deconvolution process.

The background to my skepticism began in the London School of Economics, where I attended an excellent series of lectures by Professor John Watkins on the philosophy of science. This introduced me to the works of Thomas Kuhn and Sir Karl Popper, whose chair Watkins held after Popper retired from L.S.E. As an aspiring scientist, I found much to agree with in what both Kuhn and Popper wrote; but as there was, and still is, a raging controversy between the philosophies of Kuhn and Popper (among others), I realized that I should try to understand why it was not considered possible to be in agreement with them both. I finally concluded that I had to agree with Popper's critical attitude and his famous criterion of demarcation between science and non-science.

Against this background, in 1976 I received what were, for me, revelations. First, Trevor Jowitt explained to me what had always been common practice in demultiplexing seismic data: a spherical divergence correction was applied to the demultiplexed data before the demultiplexed tape was written. I realized that this was bound to distort the wavelet. Second, Bill Lerwill told me what every experienced field geophysicist knows about seismic data: the resolution of the data depends far more on the data acquisition parameters than on the sophistication of the computer processing. At that point, I was still trying to argue that we should apply Paul Stoffa's version of homomorphic deconvolution to our data, instead of the more conventional methods. As far as Bill Lerwill was concerned, what I was talking about was fundamentally irrelevant. He proved this to me with the results of some simple experiments. I was convinced.

I became more critical of some of the steps in normal deconvolution and benefited from discussions with many friends, including Dave Brown, Roger Bilham, Paul Stoffa, Clint Frasier, and Ken Larner. From these discussions I was encouraged to give seminars and prepare notes and would like to thank Andrew Stacey, Bill Huggins, Nafi Toksöz, Chris Walker, Neil Goulty, and Wim Goudswaard, all of whom gave excellent advice on how to structure and prepare my material.

In the preparation of the manuscript, I have been greatly helped by my wife, Kate Crowley, and by Else van der Meer and Monique van der Does, all of whom typed different versions of different parts of it many times. I also thank Mr. Swannik who drafted most of the diagrams. For eighteen months my students have been very helpful in telling me which steps in the argument I never explain properly and my friend and colleague, Jacob Fokkema, has tirelessly read and reread various versions of the manuscript and checked and corrected much of the mathematics. I especially wish to thank Ken Larner for a very constructive and critical review of the manuscript, which I have since tried to improve in line with some of his suggestions, although I am aware that there is still much remaining with which Ken and many of those to whom I am indebted would disagree. The mistakes, in particular, are all my own work.

I owe a special thanks to Michael Hays of IHRDC who believed in this book before a word of it was written. He and his colleagues, Carolyn Yoder and Annette Joseph, have been a pleasure to work with and have always provided excellent professional support and exactly the right amount of encouragement at the critical moments.

I now know that the writing of a book can easily become a task from which no pleasure is derived. At no point, however, did the writing of this book develop such a character. It was prevented from doing so by my wife, Kate, who, though she encouraged me throughout and has always been willing to read and comment constructively on many parts of the book, was determined that I should still maintain *some* sense of perspective.

1. SPIKING DECONVOLUTION

1.1. The Concept of Spiking Deconvolution

In figure 1.1 we illustrate a simple view of a single-channel seismic reflection system. The seismic wave generated by the source travels down to the deep reflector and bounces back toward the surface, where the surface motion is recorded by the receiver. A number of shallow reflectors near the surface causes the primary wave packet traveling outward from the source to be multiply reflected. The seismic signal that arrives at the reflector is therefore the original wave packet followed by a tail of multiply reflected wavelets. On the way back to the surface, this signal again must pass through the shallow layered sequence, which lengthens the signal still further.

Thus the echo from this single reflector appears at the receiver as a rather complicated signal. If we measure the returning signal as a function of time τ measured relative to the onset of the arrival of the echo, we may characterize the shape of this echo as $s(\tau)$. If recording time t is measured from the shot instant and t_i is the travel time of the primary reflected sound wave from the source to the ith reflector and back to the receiver, then

$$\tau = t - t_i, \tag{1.1}$$

as shown in figure 1.2.

Of course, this particular echo is simply one of an infinite sequence of primary and multiply reflected echoes from deep reflectors. The complete sequence of echoes is characterized by the travel times $t_1, t_2, t_3, \ldots,$ and so on, and by the reflection amplitudes $g_1, g_2, g_3, \ldots,$ and so on.

Thus the sequence of echoes could be described by the function $g(t)$, such that

$$g_1 = g(t_1),$$

$$g_2 = g(t_2), \text{ etc.} \tag{1.2}$$

If every echo were received as a perfect impulse $\delta(\tau) = \delta(t - t_i)$ (see Appendix), the received seismogram would be $g(t)$, which we can express as

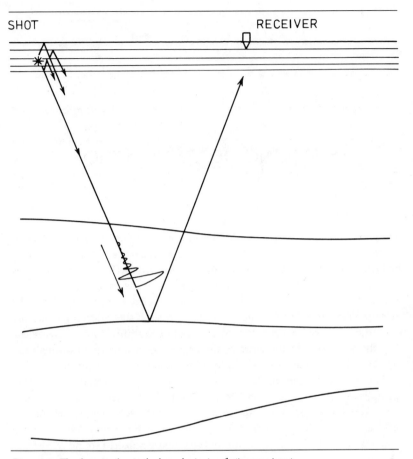

Figure 1.1. The elements of a single-channel seismic reflection experiment.

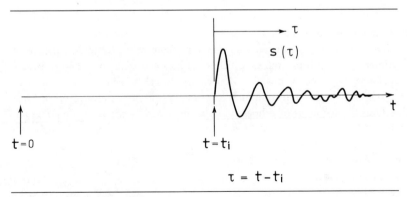

Figure 1.2. The delayed wavelet $s(\tau)$.

$$g(t) = \sum_{i=1}^{\infty} g(t_i)\,\delta(t - t_i). \tag{1.3}$$

Instead of the impulse $\delta(\tau)$, however, we receive a wavelet $s(\tau)$ for every echo $g(t_i)$. The received seismogram $x(t)$ is thus

$$x(t) = \sum_{i=1}^{\infty} g(t_i)s(t - t_i). \tag{1.4}$$

This equation is a *convolution* (see Appendix). If we think of $g(t)$ as a continuous function of time, we may write the summation as an integral:

$$x(t) = \int_0^t s(\tau)g(t - \tau)d\tau$$

$$= \int_0^t g(\tau)s(t - \tau)d\tau. \tag{1.5}$$

Alternatively, if we think of these functions of time as discretely sampled such that they form sequences at evenly spaced time intervals Δt, then the convolution can be written as

$$x_t = \sum_{\tau=0}^{t} s_\tau g_{t-\tau}$$

$$= \sum_{\tau=0}^{t} g_\tau s_{t-\tau}. \tag{1.6}$$

Equations (1.5) and (1.6) may be written in the more compact notation

$$x(t) = s(t) * g(t) = g(t) * s(t), \tag{1.7}$$

where the asterisk denotes convolution. If the functions are continuous, we imply the integral formulation (eq. 1.5); if the functions are discretely sampled, we imply the summation formulation (eq. 1.6).

It is unlikely that the echoes $g(t_i)$ are all separated sufficiently that the received wavelets do not overlap. In fact, the seismogram $x(t)$ consists of a whole sequence of overlapping wavelets. The geophysicist is much more interested in the *sequence* $g(t)$ than in the *wavelet* $s(t)$. Ideally, he or she would like to receive the sequence $g(t)$ uncontaminated by the wavelet $s(t)$. What is required to overcome this difficulty is a device that recognizes the wavelet $s(t)$ such that, whenever a new wavelet $s(t)$ arrives, the device signals its arrival by an impulse $\delta(t)$.

This device is a filter $f(t)$ whose output response is $\delta(t)$ whenever its input is $s(t)$. Thus, mathematically, we wish to define $f(t)$ such that

$$f(t) * s(t) = \delta(t),\tag{1.8}$$

as illustrated in figure 1.3. If we now pass the data $x(t)$ through this filter, we will perform the following operation:

$$\begin{aligned}
f(t) * x(t) &= f(t) * s(t) * g(t) \\
&= \delta(t) * g(t) \\
&= g(t).
\end{aligned}\tag{1.9}$$

So the filter *undoes* the harm done by the shallow layering and the consequent convolution of the echo sequence $g(t)$ with the long wavelet $s(t)$. That is, the filter deconvolves the data to remove the wavelet and reveal the echo sequence $g(t)$. These convolutional processes are illustrated in figure 1.4.

Obviously we need to know how to design this filter. The most widely used filter is the least-squares Wiener filter, described in the next section.

1.2. The Wiener Filter

A general least-squares filter model involves the three sequences illustrated in figure 1.5—namely, the input signal x_t, the desired output signal z_t, and the actual output signal y_t. The basic problem is to find the values of the filter coefficients $(f_0, f_1, \ldots, f_n)$. Wiener's (1947) solution to this problem rests on three main assumptions that determine the range of application of the results (Robinson 1967, p. 288):

1. The time series representing the input x_t and the desired output z_t can be either transient signals with finite *energy*, or infinitely long *stationary* time series with finite *power*. A stationary time series is a time series with statistical properties that do not change with time (see Appendix).
2. The approximation criterion is taken to be the mean-square error between the desired output z_t and the actual output y_t. This means that we determine the filter operator $(f_0, f_1, f_2, \ldots, f_n)$ in such a way as to minimize the mean-square error

$$I = E\{(z_t - y_t{}^2\} = E\{\epsilon_t^2\},\tag{1.10}$$

Where $\epsilon_t = z_t - y_t$ is the error series. The notation $E\{\cdot\}$ denotes the *expectation* or average of the quantity within the braces. For a transient signal this average is a sum of squares:

$$E\{\epsilon_t{}^2\} = \sum_{t=0}^{\infty} \epsilon_t{}^2;\tag{1.11}$$

that is, it is an *energy* average. For a stationary time series it is the limit of the sum of the squares divided by the elapsed time as time becomes infinite:

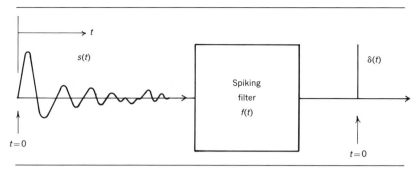

Figure 1.3. Spiking deconvolution.

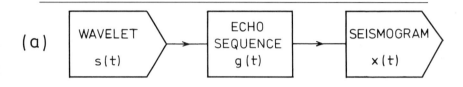

(a)

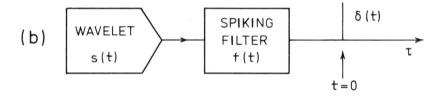

(b)

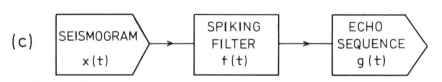

(c)

Figure 1.4. (a) Noise-free one-dimensional convolutional model of the seismogram. (b) Spiking deconvolution of the seismic wavelet. (c) Deconvolution of the seismogram.

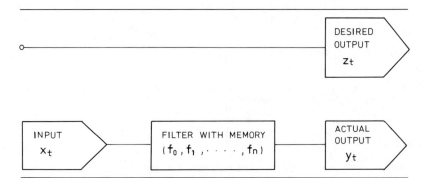

Figure 1.5. Elements in the design of a Wiener filter.

$$E\{\epsilon_t{}^2\} = \lim_{T \to \infty} \frac{1}{2T+1} \sum_{t=-T}^{T} \epsilon_t{}^2 \tag{1.12}$$

that is, it is a *power* average. In practice we never have infinitely long time series; every measured data sequence has a beginning and an end even if the measured process is of infinite duration. For stationary time series an estimate of the value of $E\{\epsilon_t{}^2\}$ can be made from a finite number of samples $2T+1$ provided this finite length sequence can be considered representative of the whole infinitely long sequence. Of course we can not know whether this is true. We can only assume that the rest of the sequence, before and after our measured sequence, has the same statistical properties as the measurement. In seismic reflection we know this assumption is false because there is always a beginning, if not an end, to the seismic reflection process initiated by the firing of the shot. Whether we have transient signals or stationary time series, the mean square error I is always a sum of squares, either equation (1.11) or an estimate of (1.12), the exact form depending on our assumptions about the kind of signals we are dealing with.

3. The filter is assumed to perform a linear operation on the available data, which are the past and present values of the input time series—that is, x_τ, where

$$-\infty \le \tau \le t, \tag{1.11}$$

and t is present time. A filter with a memory function $(f_0, f_1, \ldots)$ but no anticipation function $(f_{-1}, f_{-2}, \ldots)$ performs a linear operation on x_τ over just this range. In practice we also wish to restrict ourselves to a filter of finite length, say $(f_0, f_1, f_2, \ldots, f_n)$.

The actual output y_t is given by the convolution of the input x_t with the filter f_t:

$$y_t = \sum_{p=0}^{n} f_p x_{t-p}. \tag{1.13}$$

Substitution of y_t from equation (1.13) into equation (1.10) yields an expression for the error energy I in terms of the input x_t, the desired output z_t, and the (as yet unknown) filter coefficients $(f_0, f_1, \ldots, f_n)$.

$$I = \sum_t \left(z_t - \sum_{p=0}^{n} f_p x_{t-p} \right)^2, \tag{1.14}$$

where the form of the summation over t depends on our assumptions about the signal, as described above. Minimization of I is achieved by requiring that:

$$\frac{\partial I}{\partial f_j} = 0, \qquad j = 0, 1, \ldots, n$$

$$= \sum_t 2(z_t - \sum_{p=0}^{n} f_p x_{t-p})(-x_{t-j}) = 0, \qquad j = 0, 1, \ldots, n. \qquad (1.15)$$

Therefore

$$-\sum_t z_t x_{t-j} + \sum_t (\sum_{p=0}^{n} f_p x_{t-p}) x_{t-j} = 0, \qquad j = 0, 1, \ldots, n \qquad (1.16)$$

or

$$\sum_{p=0}^{n} f_p \sum_t x_{t-p}\, x_{t-j} = \sum_t z_t\, x_{t-j}, \qquad j = 0, 1, \ldots, n. \qquad (1.17)$$

We note that the summations on t give the autocorrelation of the input signal and the cross correlation of the input with the desired output (see Appendix):

$$\phi_{xx}(\tau) = \sum_t x_t\, x_{t-\tau}, \qquad (1.18)$$

$$\phi_{zx}(\tau) = \sum_t z_t\, x_{t-\tau} \qquad (1.19)$$

where, in both correlation functions, we have followed the notation of Robinson (1967), putting the time shift in parentheses. Equations (1.17) may now be written as

$$\sum_{p=0}^{n} f_p\, \phi_{xx}\,(j-p) = \phi_{zx}(j), \qquad j = 0, 1, \ldots, n, \qquad (1.20)$$

which are known as the "normal equations" described by Levinson (1947). If the input signal x_t is real, its autocorrelation function is perfectly symmetrical; that is

$$\phi_{xx}(\tau) = \phi_{xx}\,(-\tau). \qquad (1.21)$$

A fast method for solving the normal equations, devised by Levinson, relies on this symmetry for its success. Levinson's algorithm forms the basis of many deconvolution programs in seismic data processing packages.

In fact, the solution of the normal equations (1.20) for the least-squares Wiener filter f_p is a standard procedure once the two functions $\phi_{xx}(\tau)$ and $\phi_{zx}(\tau)$ are specified. The essence of the design of a Wiener filter is thus to define the autocorrelation $\phi_{xx}(\tau)$ of the input signal x_t and the cross correlation $\phi_{zx}(\tau)$ between the desired output z_t and the input x_t.

1.3. The Conventional Spiking Filter

More will be said about Wiener filters in later chapters. Right now, in our enthusiasm to solve the problem, it is sufficient to know that Wiener's design method leading to the normal equations exists together with Levinson's fast algorithm for solving the equations. All we need to do is apply the method to our problem of designing a spiking filter.

Clearly the immediate job is to find the filter $(f_0, f_1, \ldots, f_n)$ that will turn our wavelet $s(t)$ into a spike $\delta(t)$. That is, our desired output is the sequence

$$z_t = 1, 0, 0, 0, \ldots, \tag{1.22}$$

and our input should be the wavelet sequence

$$s_t = s_0, s_1, s_2, \ldots. \tag{1.23}$$

In order to find the filter, we need to know the left-hand-side autocorrelation coefficients of the input $\phi_{ss}(\tau)$, and the right-hand-side cross-correlation coefficients $\phi_{zs}(\tau)$.

Let us deal first with the right-hand-side coefficients, which are given by the equation

$$\phi_{zs}(\tau) = \sum_{t=\tau}^{\infty} z_t \, s_{t-\tau}, \tag{1.24}$$

and have been obtained directly from equation (1.19) by substituting s_t for the input wavelet which is assumed to be a transient (see Appendix). If we now calculate these coefficients, we find

$$\phi_{zs}(0) = s_0$$

$$\phi_{zs}(1) = 0$$

$$\vdots \qquad \vdots$$

$$\phi_{zs}(n) = 0. \tag{1.25}$$

That is, only the first coefficient on the right-hand side has a value. All the rest are zero, as illustrated in figure 1.6.

Now let us consider the left-hand-side coefficients. We need to know the autocorrelation of s_t. In fact, we do not know s_t; all we know is x_t, the convolution of s_t and g_t. Suppose we find the autocorrelation of x_t. What do we get?

It can be shown (and we will show it in chapter 3.) that the autocorrelation of x_t is equal to the autocorrelation of s_t convolved with the autocorrelation of g_t:

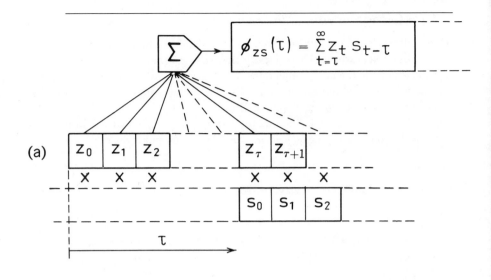

(a)

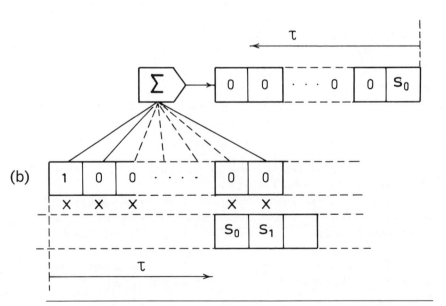

(b)

Figure 1.6. (a) The cross-correlation function $\phi_{zs}(\tau)$. (b) Cross correlation of a spike with a wavelet.

$$\phi_{xx}(\tau) = \phi_{ss}(\tau) * \phi_{gg}(\tau). \tag{1.26}$$

Now if $\phi_{gg}(\tau)$ is a sharply peaked function with very small side-lobes—preferably something like $\delta(\tau)$—the $\phi_{xx}(\tau)$ is very similar to $\phi_{ss}(\tau)$. It turns out that if the sequence of echoes $g(t)$ is white, random, and stationary with zero mean, then

$$\phi_{gg}(\tau) = \sigma^2 \delta(\tau), \tag{1.27}$$

where σ^2 is the variance of $g(t)$ (see Appendix). That is, in this case

$$\phi_{xx}(\tau) = \phi_{ss}(\tau) * \sigma^2 \delta(\tau)$$
$$= \sigma^2 \phi_{ss}(\tau). \tag{1.28}$$

In other words, if we assume that g(t) is white, random, and stationary with zero mean, we can estimate the autocorrelation coefficients $\phi_{ss}(\tau)$ from the data x_t, apart from a scale factor σ^2. Let us now write down the normal equations we must solve, from equations (1.20):

$$\sum_{p=0}^{n} f_p \, \phi_{ss} \, (p - \tau) = \phi_{zs}(\tau), \tau = 0, 1, \ldots, n. \tag{1.29}$$

We can substitute for $\phi_{ss}(\tau)$ from equation (1.28) and can substitute for $\phi_{zs}(\tau)$ from equations (1.25) to write equations (1.29) in the matrix form

$$\frac{1}{\sigma^2} \begin{bmatrix} \phi_{xx}(0) & \phi_{xx}(1) & \cdots & \phi_{xx}(n) \\ \phi_{xx}(-1) & \phi_{xx}(0) & \cdots & \phi_{xx}(n-1) \\ \vdots & \vdots & \vdots & \vdots \\ \phi_{xx}(-n) & \phi_{xx}(-n+1) & \cdots & \phi_{xx}(0) \end{bmatrix} \begin{bmatrix} f_0 \\ f_1 \\ \vdots \\ f_n \end{bmatrix} = \begin{bmatrix} s_0 \\ 0 \\ \vdots \\ 0 \end{bmatrix}, \tag{1.30}$$

which can be written as

$$\frac{1}{s_0 \sigma^2} \begin{bmatrix} \phi_{xx}(0) & \phi_{xx}(1) & \cdots & \phi_{xx}(n) \\ \phi_{xx}(-1) & \phi_{xx}(0) & \cdots & \phi_{xx}(n-1) \\ \vdots & \vdots & \vdots & \vdots \\ \phi_{xx}(-n) & \phi_{xx}(-n+1) & \cdots & \phi_{xx}(0) \end{bmatrix} \begin{bmatrix} f_0 \\ f_1 \\ \vdots \\ f_n \end{bmatrix} = \begin{bmatrix} 1 \\ 0 \\ \vdots \\ 0 \end{bmatrix}. \tag{1.31}$$

The scale factor $1/s_0 \sigma^2$ is unknown. Otherwise, however, we need only find the autocorrelation function $\phi_{xx}(\tau)$, which we can easily compute from the recorded data x_t using equation (1.18). Apart from the scale factor, then, we can solve equations (1.31) for the filter coefficients $(f_0, f_1, \ldots, f_n)$ using Levinson's algorithm. In fact, this unknown scale factor is not that important. It will only affect the absolute value of the spike output of the filter, but it will not alter the *shape* of the filter or the *shape* of its output. Thus the major objective of designing a filter that will recognize the wavelet $s(t)$ as input and give a spike as output has been accomplished.

Now that the problem has been solved conceptually, let us see what we have done.

1. We have said that the recorded data x_t are the convolution of s_t and g_t, neither of which we know. That is, we have *one* convolutional *equation containing* the *two unknowns* s_t and g_t. We wish to know g_t.
2. We design a filter f_t using the Wiener least squares filter design method; and, with some assumptions about the properties of g_t, we solve the equation to find g_t, as desired.

To put it another way, we have tried to solve one equation with two unknowns. By making guesses about the statistical properties of one of the unknowns, we find this unknown. How good is the answer? One purpose of this book is to show that it is impossible to answer this question. The point is that we have guessed the answer. Once we are guessing, it does not matter how sophisticated we are with our statistical assumptions: we are still guessing.

2. THE ONE-DIMENSIONAL CONVOLUTIONAL MODEL OF THE SEISMOGRAM

2.1. Introduction

Geophysicists have had much success in treating seismograms as the responses of linear systems. A seismogram is regarded as the response of the one-dimensional linear system whose input is a seismic wave. This one-dimensional linear-system approach has been especially important in the field of deconvolution.

In the literature of seismic deconvolution (e.g., Webster 1978) the one-dimensional convolutional model is taken as being well understood. The literature on this convolutional model itself is not extensive, however. Only by searching through the extensive literature on *de*convolution does one come to the realization that there are *two* convolutional models that geophysicists have in mind. Whenever there is a measurement of the seismic wavelet to be used in the design of a deconvolution filter, a *point-source* convolution model of the seismogram is implied. When there is no measurement available, some sort of statistical approach to the problem is invoked, in which case it is very helpful to think about a *plane-wave* convolutional model of the seismogram. In this chapter we discuss both these models. The point source model is discussed in some detail because it is now very often implied in source signature deconvolution, but has so far (to this author's knowledge) never been explained.

2.2. Linearity, Isotropy, and the Acoustic Approximation

An elastic medium will transmit sound waves. For small-amplitude sound waves the transmission is via linear elastic deformation of the medium; that is, the stress is proportional to the strain (Hooke's Law). Deviations from linearity occur at high frequencies or when the amplitude of the transmitted wave is large. Any nonlinear effects of this sort, such as frequency-dependent absorption, are ignored in the convolutional model. It is the *linearity* of the propagation of elastic waves in rocks that allows the seismologist to get so far in his analysis of seismic records.

For the purposes of the one-dimensional model of the seismogram, it is also important to assume that the rocks are *isotropic*. That is, the elastic behavior is not a function of the direction of propagation of the waves. In many sedimentary rocks there is a preferred orientation of the particles within the rock matrix that gives a bias to elastic properties such that sound waves will travel faster in one direction than in another. This is called *anisotropy*. In both the point-source and plane-wave convolutional models, the possible existence of anisotropy is ignored. In such homogeneous linear elastic isotropic media two kinds of waves may propagate: *compressional* and *shear* waves. Most seismic sources generate their elastic energy predominantly as compressional waves. As these waves propagate through the elastic layers they may be partly converted to shear waves at interfaces between two layers, both on reflection and on transmission, as illustrated in figure 2.1. Similarly, the converted shear waves will also be partly converted back to compressional waves at interfaces between two media. At each interface, the fraction of energy which is converted from compressional to shear, or vice versa, depends on the angle of incidence of the wave (see Aki and Richards 1980, ch. 5). At normal incidence there is no conversion, and conversions do not become important until the angle of incidence is large. Model studies of this problem (e.g., Hughes and Kennett 1983) agree that conversions can be neglected for small angles of incidence and thus, for most purposes, only compressional waves need be considered. This is known as the *acoustic approximation,* because acoustic media are fluids which do not support shear stresses and in which shear waves therefore do not propagate. In both the point-source and plane-wave convolutional models, the acoustic approximation is made. Therefore an incident wave is converted into a single reflected wave and a single transmitted wave at each interface.

2.3. Point-Source Convolutional Model

SIMPLE TRANSMISSION. Consider first a point source and point receiver buried in an infinite homogeneous isotropic elastic medium as shown in figure 2.2(a). Close to a dynamite source there is a small nonlinear region. Outside this region, at some distance that we call the *elastic radius,* the compressional wave propagates according to the laws of elasticity. The source of the elastic waves may be thought of as acting over a sphere rather than at a point. For the purposes of this chapter, we do not need to consider the complications that arise from the existence of the nonlinear zone at the source. For the initial description of the point-source model, it is adequate to treat the waves as if they originated at a point.

Let the source emit an impulsive compressional wave at time $t = 0$. The wave will propagate out from the point source with spherical symmetry, the wavefront traveling at velocity c, the velocity of compressional waves in the medium. If the distance between the source and receiver is r, then the wave will arrive at the receiver at time

$$t_1 = r/c. \tag{2.1}$$

Since the process of elastic wave propagation in the medium is linear, the shape of the wave remains the same throughout its propagation through the medium. That is, it remains an impulse. The receiver thus receives an impulse at time t_1 of amplitude a_1 reduced by spherical divergence. Thereafter nothing else happens at the receiver; the wave has gone past and will not return.

The response of this system to an impulse at the source at time $t = 0$ is simply a delayed impulse of reduced amplitude, as depicted in figure 2.2(b), where we have called the system response $a(t)$

$$\begin{aligned} a(t) &= a_1, & t &= t_1 \\ &= 0, & t &\neq t_1. \end{aligned} \tag{2.2}$$

Consider now that the source and receiver remain in the same positions, but that the source emits a more complicated wave $s(t)$, which begins at time $t = 0$ and extends for some short duration as depicted in figure 2.3. Once again, since our source is a point, this wave will propagate outward with spherical symmetry and will travel with velocity c through the medium. It will arrive at the receiver, unchanged in shape, at time $t = t_1$ as shown in figure 2.4. Its amplitude will be scaled by the factor a_1, just like the impulse, because this factor is a propagation effect caused by the *divergence of the wavefront* over the distance r. If we call $x(t)$ the receiver response of our system to the input $s(t)$ at the source, we can deduce that

$$\begin{aligned} x(t) &= a_1 s(t - t_1) \\ &= a(t_1) s(t - t_1). \end{aligned} \tag{2.3}$$

The wavelet that arrives has exactly the same shape as the one that started out at the source. The only things that have happened to it are that it has been delayed and its amplitude has been reduced. If we define a parameter τ as

$$\tau = t - t_1, \tag{2.4}$$

this will specify a point on the wave independent of the time delay t_1, as shown in figure 2.5. That is, $s(\tau)$ is our wavelet of constant shape. We note that $s(\tau)$ is zero when τ is negative, because this corresponds to the *causality*

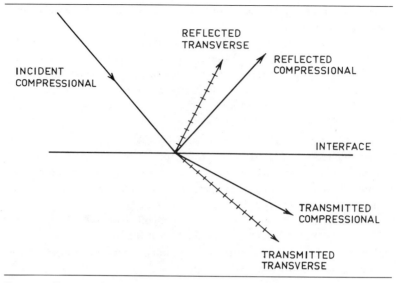

Figure 2.1. Conversion of waves at an elastic interface.

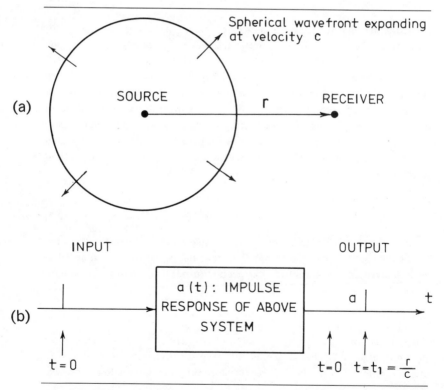

Figure 2.2. (a) Point source and point receiver in infinite homogeneous isotropic elastic medium. (b) Linear system representation of (a).

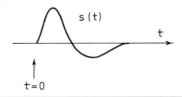

Figure 2.3 Source wavelet s(t).

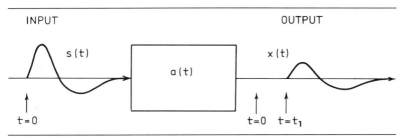

Figure 2.4. Linear system representation of figure 2.2(a) with s(t) as source wavelet.

s (t)

s (t)   delayed
        by time t_1

s (τ) specifying
        shape of wave
        regardless of
        delay

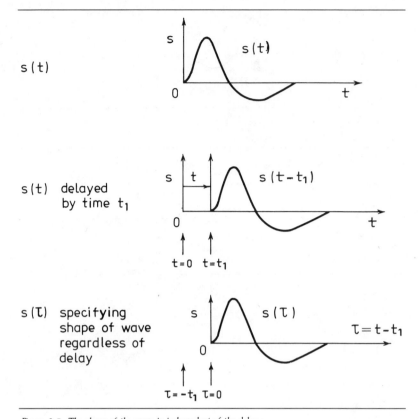

Figure 2.5. The shape of the wave is independent of the delay.

condition that the source did not generate a wave before $t = 0$. If we substitute for τ from equation (2.4) into equation (2.3), we have

$$
\begin{aligned}
x(t) &= a(t_1)s(\tau) \\
&= a(t - \tau)s(\tau).
\end{aligned}
\tag{2.5}
$$

Equation (2.5) says that our output $x(t)$ is equal to a wavelet of shape $s(\tau)$, depicted in figure 2.5, scaled in amplitude by a factor a_1, and delayed by a time t_1.

SINGLE-REFLECTOR MODEL. Let us suppose that we have a slightly more complicated system: a homogeneous linear isotropic elastic half-space above a rigid reflector as shown in figure 2.6(a). Once again we have a point source and point receiver buried in the medium and separated by a distance r_1. If our source emits an impulse at time $t = 0$, our receiver will receive an impulse at time t_1 where

$$
t_1 = r_1/c,
\tag{2.6}
$$

corresponding to the transmission of the direct wave from the source; it will also receive a reflection at time t_2, appearing to come from the image of the source in the reflector

$$
t_2 = r_2/c.
\tag{2.7}
$$

The amplitudes of these two impulses are controlled by the propagation of the spherical wave in the half-space, and by the laws of reflection. If the amplitudes of these two impulses are b_1 and b_2, we can depict this situation as a linear system with impulse response $b(t)$, as shown in figure 2.6(b), such that

$$
\begin{aligned}
b(t) &= b_1, & t &= t_1 \\
&= b_2, & t &= t_2 \\
&= 0, & t_1 &\neq t \neq t_2.
\end{aligned}
\tag{2.8}
$$

The reflection at time t_2 obviously arrives later than the direct wave at time t_1, because the wave travels at constant velocity c and the reflected wave has further to travel than the direct wave.

If our source now emits a wave $s(t)$ at time $t = 0$, as before, then we would receive two arrivals at the receiver, delayed and scaled by the same amounts as for the impulsive wave, as shown in figure 2.7. If our signal received at the receiver is $x(t)$, we may describe this as

$$
\begin{aligned}
x(t) &= b_1\, s(t - t_1) + b_2 s(t - t_2) \\
&= b(t_1)\, s(t - t_1) + b(t_2)s(t - t_2) \\
&= \sum_{i=1}^{2} b(t_i)s(t - t_i).
\end{aligned}
\tag{2.9}
$$

If we wish to show more explicitly that the wavelet has the same shape, independent of the travel time or travel path, we may do this using the same device as before. That is, we define the parameter

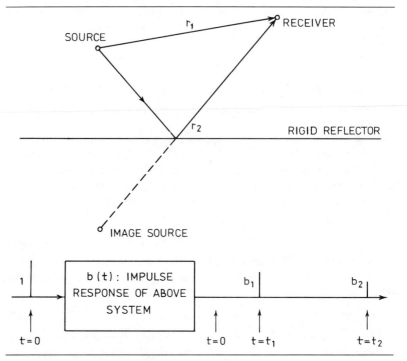

Figure 2.6. (a) Point source and point receiver in homogeneous isotropic elastic half-space above a rigid reflector. (b) Linear system representation of (a).

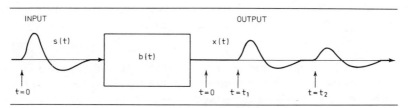

Figure 2.7. Linear system representation of 2.6(a) with s(t) as source wavelet.

$$\tau_i = t - t_i, \qquad i = 1 \text{ or } 2. \tag{2.10}$$

So τ_i is measured relative to the arrival of the onset of the wave at time t_i. From equations (2.9) and (2.10) we can rewrite our receiver response $x(t)$ as

$$
\begin{aligned}
x(t) &= b_1 s(\tau_1) + b_2 s(\tau_2) \\
&= b(t_1) s(\tau_1) + b(t_2) s(\tau_2) \\
&= \sum_{i=1}^{2} b(t - \tau_i) s(\tau_i).
\end{aligned}
\tag{2.11}
$$

If our rigid reflector is closer to our source and receiver than we have shown it, such that the travel time difference $t_2 - t_1$ becomes shorter, it is possible that the arrival of the reflection will overlap the tail of the direct wave. It is a consequence of the theory of linear elastic media that pressure waves superpose, provided the resultant amplitude is still within the linear elastic stress range. Therefore, even if the arriving wavelets overlap, we are still correct to write down our response as a straight summation, as we have done in equations (2.9) and (2.11).

COMPLETE MODEL. Suppose we have a point source and point receiver buried in an infinite sequence of layers of homogeneous elastic media, each characterized by density ρ_i, and compressional wave velocity c_i, as shown in figure 2.8. On the emission of an impulse by the point source at time $t = 0$, a spherical propagating wave will be generated. At each boundary between two layers, the wave will be reflected and refracted according to the laws of reflection and refraction. Since some energy is always transmitted across the boundary, the receiver will receive an infinitely long sequence of impulses at times

$$t_1, t_2 \ldots .$$

The amplitudes of these impulses will gradually become infinitely small as the wave travels farther and farther and becomes weaker and weaker in intensity. Let us call the response of this system $g(t)$, such that

$$g_1 = g(t_1),$$

$$g_2 = g(t_2), \text{ etc.} \tag{2.12}$$

As before, the shape of the wave is unchanged by its propagation through the series of elastic layers. The amplitudes and arrival times of the sequence of impulses $g(t)$ are controlled by the geometry of the source and receiver and by the paths that the waves take between them.

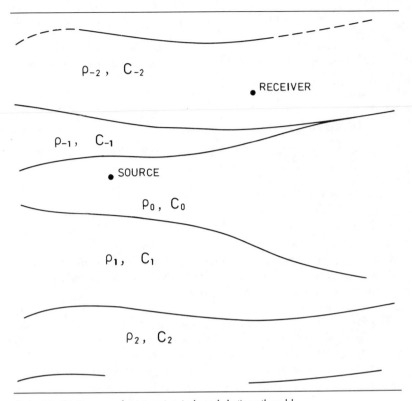

ρ_{-2}, C_{-2}

RECEIVER

ρ_{-1}, C_{-1}

SOURCE

ρ_0, C_0

ρ_1, C_1

ρ_2, C_2

Figure 2.8. Point source and point receiver in layered elastic earth model.

If we now consider that our point source emits a spherical wave $s(t)$, we will expect to receive a sequence $x(t)$ at our receiver, where

$$
\begin{aligned}
x(t) &= g_1 s(t - t_1) + g_2 s(t - t_2) + \ldots \\
&= g(t_1) s(t - t_1) + g(t_2) s(t - t_2) + \ldots \\
&= \sum_{i=1}^{\infty} g(t_i) s(t - t_i),
\end{aligned}
\tag{2.13}
$$

where the causality condition is that $s(t - t_i)$ is zero when $(t - t_i)$ is negative, thus ensuring that echoes which arrive after time t are not included.

We now use the same device as before to define a point on the wavelet relative to its arrival time t_i:

$$
\tau_i = t - t_i,
\tag{2.14}
$$

and we may rewrite equation (2.13) as

$$
x(t) = \sum_{i=1}^{\infty} g(t - \tau_i) \cdot s(\tau_i).
\tag{2.15}
$$

This summation is a convolution. The summations in equations (2.9), (2.11) and (2.13) are also convolutions. The convolution in equations (2.13) and (2.15) may be written as integrals, provided we think of $g(t)$ as a response that varies continuously with time t. This is in contrast to our previous concept of a series of impulses at discrete times. In the continuous case we may write equation (2.15) as

$$
x(t) = \int_0^{\infty} g(t - \tau) s(\tau) d\tau.
\tag{2.16}
$$

Similarly, equation (2.13) may be written as

$$
x(t) = \int_0^{\infty} g(\tau) s(t - \tau) d\tau.
\tag{2.17}
$$

Equations (2.16) and (2.17) are equivalent expressions for the description of $x(t)$. If we now consider that $x(t), g(t)$, and $s(t)$ are sampled at regular discrete intervals Δt such that the continuous functions may be reconstructed from the discrete functions, we may write

$$
x_t = x_0, x_1, x_2, \ldots,
$$
$$
g_t = g_0, g_1, g_2, \ldots,
\tag{2.18}
$$
$$
s_t = s_0, s_1, s_2, \ldots,
$$

where the three sequences must be related as follows:

$$x_t = \sum_{\tau=0}^{\infty} g_{t-\tau} s_\tau$$

$$= \sum_{\tau=0}^{\infty} g_\tau s_{t-\tau}. \tag{2.19}$$

In the derivation of this point-source one-dimensional convolutional model of the seismogram, we have tried to show that every wavelet $s(\tau)$ that arrives at the receiver corresponds to a distinct arrival within the impulse response $g(t)$. In order that the received seismogram is a convolution, it is essential that the received wavelet is the same shape, regardless of the arrival time of the wavelet or the direction from which it comes. The received wavelets will have the same shape only if the source emits a wave that has the same shape in all directions.

2.4. Problems of Source and Receiver Directivity

If the source has a *directivity pattern* such that the shape of the wavelet transmitted is a function of the direction, then the receiver will receive a seismogram that is *not* a convolution of $s(t)$ with $g(t)$; for the convolutional model implies that the shape of the wave $s(\tau)$ is the same for every arrival within the impulse response $g(t)$.

If the direction of the source wave is specified by the two angles θ_s and φ_s relative to the Cartesian coordinate system shown in figure 2.9, where the origin of the system is at the source, then we may specify the shape of the wave at the source as $s(\theta_s,\varphi_s,t)$. The receiver will receive a sum of arrivals corresponding to the same sequence of arrivals of the impulse response $g(t)$, but also depending on the direction (θ_s,φ_s) in which the wave began its travel path on its way from the source to the receiver.

In fact, the impulse response $g(t)$ contains *all* the arrivals. There will be a sub-set $g(\theta_s,\varphi_s,t)$ of these which consists only of those arrivals that originated at the source in the direction (θ_s,φ_s). For this sub-set of arrivals, the shape of the wave is unchanged as it propagates through the media and the resulting response $x(\theta_s,\varphi_s,t)$ is therefore the convolution

$$x(\theta_s,\varphi_s,t) = \int_0^t g(\theta_s,\varphi_s,t-\tau)s(\theta_s,\varphi_s,\tau)d\tau. \tag{2.20}$$

The total response $x(t)$ is the sum of all such sub-set responses; that is, it is an integration over all possible source directions:

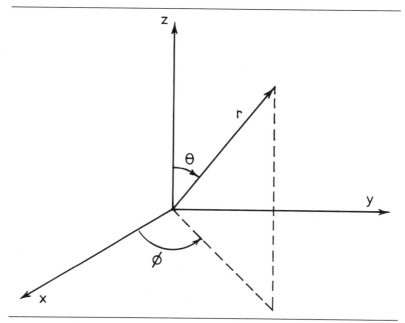

Figure 2.9. Cartesian coordinates and direction angles θ and φ.

$$x(t) = \int_0^{2\pi} \int_0^{2\pi} x(\theta_s,\varphi_s,t)d\theta_s d\varphi_s$$

$$= \int_0^{2\pi} \int_0^{2\pi} \int_0^{\infty} g(\theta_s,\varphi_s,t-\tau)s\,(\theta_s,\varphi_s,\tau)d\tau d\theta_s d\varphi_s. \tag{2.21}$$

A similar effect will occur at the receiver if it has a directivity pattern. To see how this further complicates the problem, consider the response $x(\theta_s, \varphi_s,t)$ of equation (2.20). This contains all the arrivals at the receiver that originated at the source in the direction (θ_s,φ_s). There are no constraints on the direction from which these arrivals may arrive at the receiver. They may arrive from many different directions, depending on the structure of the geology. If we select one arrival direction (θ_r,φ_r) (see figure 2.10) at the receiver, we may denote the sub-set of arrivals that corresponds to *both* the chosen source direction *and* the chosen receiver direction as $g(\theta_s,\varphi_s,\theta_r, \varphi_r,t)$.

Of the sound wave which leaves the source in direction (θ_s,φ_s) the part that arrives at the receiver in the direction (θ_r,φ_r) is the convolution of $s(\theta_s,\varphi_s,t)$ with this sub-set of arrivals:

$$x(\theta_s,\varphi_s,\theta_r,\varphi_r,t) = \int_0^t g(\theta_s,\varphi_s,\theta_r,\varphi_r,t-\tau)\, s(\theta_s,\varphi_s,\tau)d\tau. \tag{2.22}$$

If we let the receiver response in this direction be $r(\theta_r,\varphi_r,t)$—that is, time-dependent as well as direction-dependent—we find the response $y(\theta_s, \varphi_s,\theta_r,\varphi_r,t)$ of the receiver to this selected sub-set of waves as the convolution:

$$\begin{aligned} y(\theta_s,\varphi_s,\theta_r,\varphi_r,t) &= \int_0^t x(\theta_s,\varphi_s,\theta_r,\varphi_r,\ t-u)r(\theta_r,\varphi_r,u)du \\ &= x(\theta_s,\varphi_s,\theta_r,\varphi_r,t)*r(\theta_r,\varphi_r,u) \\ &= s(\theta_s,\varphi_s,t)*g(\theta_s,\varphi_s,\theta_r,\varphi_r,t)*r(\theta_r,\varphi_r,t). \end{aligned} \tag{2.23}$$

The total response $y(t)$ of the receiver is the sum of all such sub-set responses; that is, it is an integration over all possible source and receiver directions:

$$y(t) = \int_0^{2\pi} \int_0^{2\pi} \int_0^{2\pi} \int_0^{2\pi} y(\theta_s,\phi_s,\theta_r,\phi_r,t)\,d\theta_s d\phi_s d\theta_r d\phi_r. \tag{2.24}$$

This is a sum of an infinite number of one-dimensional convolutions, each of which is described by equation (2.23).

The directivity pattern $r(\theta,\varphi_r,t)$ of the receiver array can usually be calculated using linear antenna theory (see, for example, Robinson 1967, chapter 5), provided the geometry of the receiver array and the sensitivities of the individual receiver elements are known. For marine hydrophone receiver

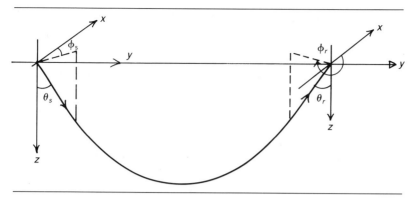

Figure 2.10. Showing source and receiver direction.

arrays this calculation is usually simple and the directivity pattern is known quite accurately. For land data, the accuracy of the calculation is usually more uncertain because variations in ground coupling of the geophones within a single geophone group cause variations in their sensitivities; it is likely therefore that every geophone group will have a slightly different directivity pattern.

The directivity pattern $s(\theta_s, \phi_s, t)$ of most seismic source arrays is usually not known. The linear antenna theory which can be applied so successfully to the receiver cannot usually be applied to the source because the essential assumption that the amplitude of the transmitted wavefield is small in the vicinity of the arrays does not apply near to the seismic source. Seismic sources generate high amplitude compressional waves and the individual elements of a seismic source are coupled via the medium to their neighbors (see Ziolkowski et al. 1982). This coupling via the medium is negligible for elements of a seismic receiver array, because the amplitudes of the received signal are so much weaker than those of the signal transmitted at the source.

The directivity pattern of either the source or receiver array is frequency-dependent: the response of the array becomes an increasingly sensitive function of direction as the frequency increases. Directivity is negligible when the size of the array is small compared with the wavelength λ, where

$$\lambda = c/f, \tag{2.25}$$

in which c is the velocity of wave propagation and f is the frequency. When the directivity is negligible the array has the same response in all directions. That is, it has spherical symmetry. Directivity becomes important when the array size is of the order of a wavelength or greater (see Stoffa and Ziolkowski 1983).

In summary, the point-source one-dimensional convolutional model of the seismogram assumes that the source and receiver array have negligible directivity over the frequency bandwidth of interest. If the source and receiver do have significant directivity functions, the one-dimensional convolutional model does not apply; the seismogram then consists of the summation of many different one-dimensional convolutions, as described by equation (2.24). If the bandwidth of the seismogram is to be preserved, the problem becomes three-dimensional, and both the source and receiver directivity pattern need to be known. It is usually much more difficult to determine the source directivity pattern than the receiver directivity pattern. If either the source or receiver directivity pattern is significant, and the one-dimensional convolutional model is applied, then it is valid only at frequencies at which directivity can be neglected; that is, at low frequencies, where the size of the array is small compared with the wavelength. *In other words, this one-dimensional convolutional model is a low-frequency approximation to the structure of the seismogram.*

To illustrate the effect of directivity, consider the case of an air-gun subarray consisting of 7 air guns spaced over a length of about 20 m, fired simultaneously at a depth of 7.5 m and a pressure of 2,000 psi (136 bars). The configuration is shown in figure 2.11. If sufficient measurements of the near-field pressure field are made, it is possible to calculate the far-field pressure field of this array of point sources using the method of Ziolkowski et al. (1982). Figure 2.12 shows how the far-field signature of the subarray of figure 2.11 varies as a function of angle to the vertical; these signatures have been calculated from near-field measurements according to the method of Ziolkowski et al. (1982). Clearly, the effect of directivity is significant, reducing the amplitude of the high-frequency first peak relative to the lower-frequency tail.

In practice, in marine seismic surveying it is now not uncommon to use arrays that are 70 m long by 70 m wide. Source directivity becomes noticeable where the source is of the same size as a wavelength. A wavelength of 70 m in water corresponds to a frequency of about 20 Hz (the speed of sound in water is about 1500 ms $^{-1}$). We would therefore expect directivity to be noticeable at about 20 Hz and to be a serious problem above about 50 Hz. In such surveys the one-dimensional convolutional model can be considered accurate only at frequencies well below 50 Hz.

2.5. Plane-Wave Model

Two basic concepts are involved in the one-dimensional plane-wave model of the seismogram: first, the earth consists of a sequence of horizontally stratified layers, and second, the source is a plane, also horizontal.

In the earth model, shown in figure 2.13, each layer is characterized by its thickness and elastic properties. Only the normal-incidence compressional waves are considered. The vertical two-way travel time of compressional waves in each layer is very often considered to be a multiple of some suitably small time Δt. The time-domain description may then be made in terms of discrete samples rather than in terms of continuous time. Figure 2.14 shows how the travel paths are related to time.

From our discussion of the one-dimensional point-source model, we know that an arrival at time, say, $t = t_1$, has an amplitude that depends on the divergence of the spherical wave; the laws of reflection and refraction, and the geometrical configuration of the source, receiver, and geology. In the plane-wave model with normally incident compressional waves, the configuration of the geology is as simple as we can make it, and the response of the system is independent of the horizontal position of the receiver. When normal-incidence plane compressional waves are considered in the context of the horizontally stratified earth model, there are no refractions. The model

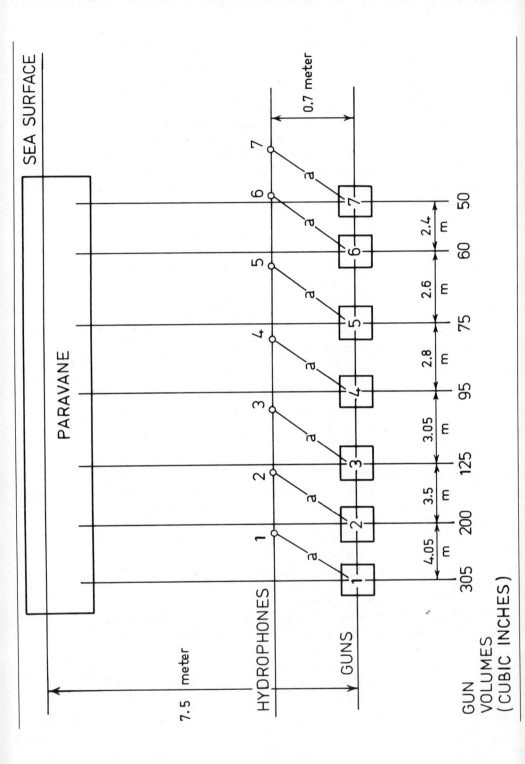

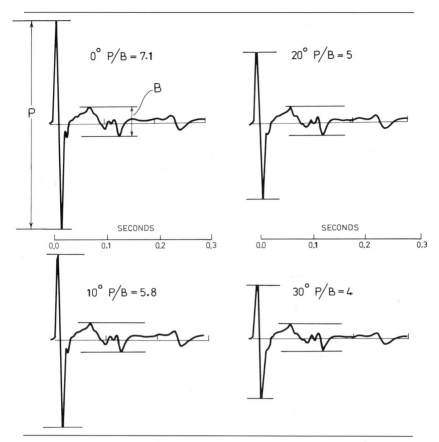

Figure 2.12. Variation of source signal with angle of incidence to vertical for air-gun subarray of figure 2.11.

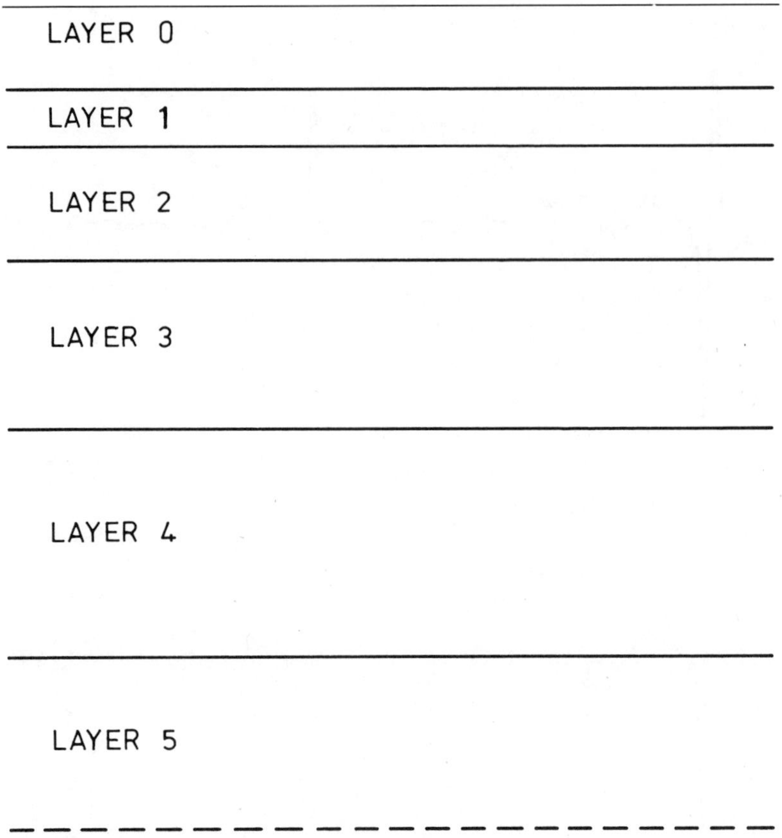

LAYER 0
LAYER 1
LAYER 2
LAYER 3
LAYER 4
LAYER 5

Figure 2.13. Horizontally stratified earth model.

TIME
0
Δt/2
2Δt/2
3Δt/2
4Δt/2
5Δt/2
Plane wave source pulse at t=0
LAYER 0
LAYER 1
LAYER 2
LAYER 3

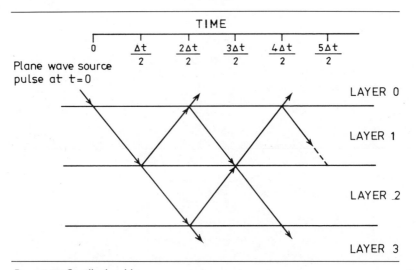

Figure 2.14. Goupilland model.

is purely one-dimensional. Only reflections are considered. At each interface there is a reflection and a transmission coefficient. The sequence of reflection coefficients is often known as the *reflectivity sequence.*

In this model an arrival at time $t = t_1 = n\Delta t$ is the sum of all arrivals, including those that are multiply reflected, for which the total two-way travel time is $n\Delta t$. There is a finite number of such arrivals. The amplitude of the sum of these arrivals depends only on the normal-incidence reflection coefficients. Thus the reflection impulse response $r(t)$ of the model can be characterized by the reflection coefficients at the interfaces between the horizontally stratified sequence of layers each with two-way travel time Δt. The reflection coefficients are of course zero whenever the elastic material on both sides of an interface is the same. This model has been considered in terms of z-transforms by many authors including Goupilland (1961) and Treitel and Robinson (1966).

If the normal-incidence plane wave applied at $t = 0$ is not an impulse but is instead the more complicated wave $s(t)$, then the resultant response $y(t)$ is the convolution of $s(t)$ and $r(t)$:

$$y(t) = s(t) * r(t). \tag{2.26}$$

The argument for using such a model is usually two-fold. First, far away from a point source the curvature of the wavefront is slight and the wavefront can be considered a plane. Second, the angle of incidence in seismic reflection is often small, so normal incidence is a reasonable assumption. Comparing the plane-wave model with the point-source model, we should note two reasons why the point-source model is superior. First, the divergence of the wavefront is usually not negligible over the range of depths from which reflections are received. Second, the plane-wave model cannot explain the differences in received seismograms that occur with variations in source-receiver separation.

If the point-source model is favored, it must be remembered that the impulse response of the earth, $g(t)$, is not the same as the reflectivity sequence, and is also not the same as $r(t)$, which is the plane-wave response.

2.6. The Problem of Reconciling Well Logs with Seismograms

Whether the computation is done in the time domain or the frequency domain, it is this plane-wave horizontally stratified earth model that is generally used in the construction of synthetic seismograms from well logs. Whenever a synthetic seismogram is compared with a real seismogram, in which geometrical divergence is an inevitable factor, some sort of correction is applied. This correction is an attempt to force a three-dimensional situation to fit a one-dimensional model.

2.7. Spherical-Divergence Correction and Wavelet Distortion

The plane-wave model is purely one-dimensional. It is possible that in some areas of the world this layer cake model of the *earth* is a very good approximation to the earth. The plane-wave approximation to the *wave* generated by the source, however, may well be wildly inaccurate, especially for shallow reflections.

No form of spherical-divergence correction applied to the recorded seismogram can ever restore the seismogram to what would have been recorded if the source had actually generated a plane wave. This is partly because the seismogram consists of primary reflections and multiples. There is some spherical-divergence correction that could be applied to the seismogram to restore the amplitudes of the primary reflections; this will consist of a time-varying function $a(t)$ that can be used to multiply the seismogram to restore the amplitudes of the reflections to their correct values. Since the seismogram also contains sequences of multiples that have traveled along paths that are different from those of the primaries that arrive at the same times, however, this function cannot be correct for the multiples. In fact, there is no simple time-varying function $a(t)$ that can correct the seismogram for the spherical divergence of all the multiples.

Furthermore, if the wavelet $s(t)$ is not an impulse, then any spherical-divergence correction $a(t)$ that is correct for the amplitudes of reflected impulses will distort the wavelet unevenly down the seismogram. We see this as follows.

Let the response of the horizontally layered earth to an impulsive *point* source be $r(t)$, and let the corresponding response to an impulsive plane wave be $\hat{r}(t)$. Then we would define the ideal spherical divergence correction $a(t)$ as

$$\hat{r}(t) = a(t)r(t). \tag{2.27}$$

As discussed above, $a(t)$ cannot exist if the responses include multiples. We must also assume that we have only normal-incidence reflections.

If the point-source wavelet is $s(t)$, we can describe our seismogram $x(t)$, with all these restrictions, as

$$x(t) = s(t) * r(t)$$
$$= \int_0^t s(\tau)r(t - \tau)d\tau. \tag{2.28}$$

If we wish to correct our seismogram for spherical divergence, we must multiply by $a(t)$:

$$\hat{x}(t) = a(t)x(t), \tag{2.29}$$

where $\hat{x}(t)$ is the simulated plane-wave response. Combining equations (2.23) and (2.24), we have

$$\hat{x}(t) = a(t)x(t) = a(t) \int_0^\infty s(\tau)r(t - \tau)d\tau$$

$$= \int_0^\infty \left\{ \frac{a(t)}{a(t - \tau)} s(\tau) \right\} \{ a(t - \tau)r(t - \tau)d\tau. \qquad (2.30)$$

In this integral there are two factors. The factor $a(t - \tau)r(t - \tau)$ is the impulse response, properly corrected for spherical divergence. The other factor,

$$\frac{a(t) \cdot s(\tau)}{a(t - \tau)},$$

is an undesired distortion of the wavelet as illustrated in figure 2.15. Furthermore, this distortion varies with the position in time of the wavelet in the seismogram. The only spherical-divergence correction that does *not* result in uneven distortion occurs when $a(t)$ happens to be of the form

$$Ae^{\alpha t},$$

for then the distortion factor becomes

$$e^{\alpha \tau}.$$

Of course, there is no reason that an exponential correction should be correct.

Let $a(t)$ be of the form bt^n. The distortion factor on $s(\tau)$ is then

$$\frac{bt^n}{b(t - \tau)^n}.$$

Thus spherical-divergence correction distorts the wavelet by increasing the amplitude of the tail. Only the exponential spherical-divergence correction applies a uniform distortion, and there is no guarantee that this form of correction or any other form is right. All other forms of spherical-divergence correction distort the wavelet unevenly down the seismogram and therefore invalidate the convolutional model.

2.8. Limitations of the One-Dimensional Linear-System Approach to the Analysis of a Seismogram

We have described how a seismogram can be thought of as the response of a linear system to an input wavelet $s(t)$ by the use of two models: the point-source model and the plane-wave model. We find that these models are not adequate to deal with a number of aspects of seismic data, including source and receiver directivity and spherical divergence.

If we wish to take spherical divergence into account, we cannot cope with the multiple reflections. Even if we happen to find the "correct" spherical-

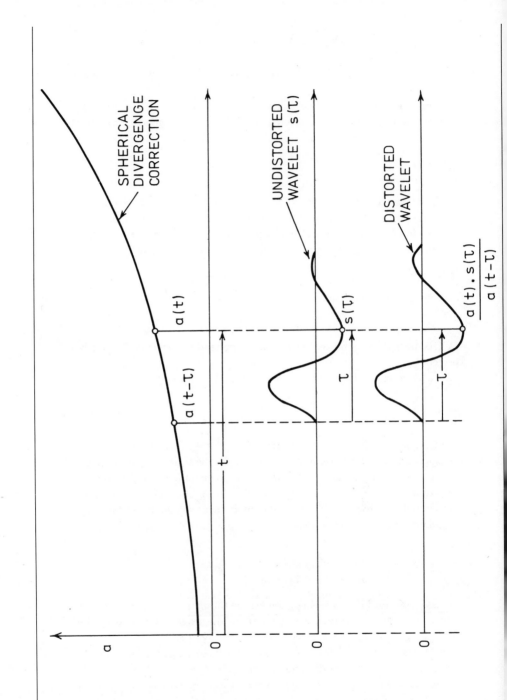

SPHERICAL
DIVERGENGE
CORRECTION

$a(t)$

$a(t-\tau)$

t

UNDISTORTED
WAVELET $s(\tau)$

$s(\tau)$

τ

DISTORTED
WAVELET

$\dfrac{a(t) \cdot s(\tau)}{a(t-\tau)}$

τ

a

0

0

0

divergence correction for the primary reflections, we are almost certain to distort the wavelet unevenly down the seismogram and thus to invalidate the convolutional model. Reconciling synthetic seismograms from well-log data with real seismograms is a real problem. The way it is usually done, using the plane-wave one-dimensional convolutional model, is clearly not correct. The problem of matching well-synthetic data with real seismograms deserves a book to itself.

3. MINIMUM PHASE AND MINIMUM DELAY

3.1. Introduction

The concept of minimum phase is central to the deconvolution process. It is not possible to appreciate the limitations of any deconvolution process without first understanding the constraints imposed by the underlying concept of minimum phase. There are many ways of defining minimum phase and they are all related, of course. This chapter aims to relate several of the most commonly used definitions to each other, and to emphasize, whenever possible, the connections between the abstract mathematics and the physical properties of the wavelet under consideration.

3.2. Causality and the Definition of Minimum Phase

We may define the concept of minimum phase in many different ways. Here we define a *minimum-phase wavelet* as a causal wavelet of finite energy whose inverse is also a causal wavelet of finite energy. A causal wavelet does not exist before time $t = 0$. Thus the wavelet $a(t)$ is causal if

$$a(t) = 0, t < 0. \tag{3.1}$$

It has finite energy if

$$E_a(\tau) = \int_0^\tau a^2(t) \, dt \tag{3.2}$$

converges as τ tends to infinity.

The *inverse* $b(t)$ of $a(t)$ is defined by the equation

$$\delta(t) = \int_{-\infty}^\infty a(\tau)b(t - \tau) \, d\tau = a(t) * b(t), \tag{3.3}$$

where $\delta(t)$ is the Dirac delta function (see Appendix). The inverse $b(t)$ is causal if

$$b(t) = 0, t < 0 \tag{3.4}$$

and has finite energy if

$$E_b(\tau) = \int_0^\tau b^2(t) \, dt \tag{3.5}$$

converges as τ tends to infinity.

Thus $a(t)$ is minimum phase if equations (3.1), (3.3), and (3.5) are satisfied and if $E_a(\tau)$ and $E_b(\tau)$ are both finite for all τ.

3.3. The z-Transform and Minimum Phase

Consider a causal wavelet $a(t)$ described by the two discrete samples

$$a_0, a_1,$$

whose z-transform $A(z)$ is (see Appendix)

$$A(z) = a_0 + a_1 z. \tag{3.6}$$

This wavelet $a(t)$ has an *inverse* $b(t)$ whose z-transform $B(z)$ is defined by

$$B(z) = \frac{1}{A(z)}. \tag{3.7}$$

We may substitute for $A(z)$ in equation (3.7) from equation (3.6) and expand in a binomial series

$$B(z) = \frac{1}{a_0} \left[1 - \frac{a_1 z}{a_0} + \left(\frac{a_1}{a_0}\right)^2 z^2 - \left(\frac{a_1}{a_0}\right)^3 z^3 + \ldots \right]. \tag{3.8}$$

The series in brackets in equation (3.8) is infinitely long. The coefficients form a convergent series, however, provided

$$\left| \frac{a_1}{a_0} \right| < 1. \tag{3.9}$$

Moreover, the sum

$$E_b(\tau) = \sum_0^\tau b_t^2 \tag{3.10}$$

converges provided the inequality (3.9) is satisfied. In this case it is possible to approximate $B(z)$ with a finite number of terms; thus an approximate causal inverse $b(t)$ of $a(t)$ exists. The exact inverse, though infinitely long, is physically realizable in the sense that its energy is finite.

If, on the other hand,

$$\left|\frac{a_1}{a_0}\right| > 1, \tag{3.11}$$

then the terms in the binomial expansion of equation (3.8) become larger and larger. The sum (3.10) *di*verges, as τ tends to infinity. That is, the energy of the inverse wavelet $b(t)$ becomes infinite. We can make an approximate inverse of finite energy by truncating the series. How good an approximation will this truncated series be? Since the infinite series diverges, the terms that have been ignored by the truncation are infinitely more important than the terms remaining in the truncated series. The approximation will therefore be extremely poor.

Alternatively, however, we can expand $B(z) = 1/A(z)$ in *de*creasing powers of z as

$$B(z) = \frac{1}{A(z)}$$

$$= (a_0 + a_1 z)^{-1}$$

$$= \frac{z^{-1}}{a_1}\left[1 - \left(\frac{a_0}{a_1}\right)z^{-1} + \left(\frac{a_0}{a_1}\right)^2 z^{-2} - \left(\frac{a_0}{a_1}\right)^3 z^{-3} + \ldots\right]. \tag{3.12}$$

Now, if the inequality (3.11) holds, the series converges and again an approximation to the exact inverse may be made with a finite number of terms. Every term in the series, however, is a negative power of z; that is, each term corresponds to a point in time *before* $t = 0$. In other words, the exact inverse of $a(t)$ is totally anti-causal. It is a linear filter $b(t)$ whose impulse response is produced before the input is applied. From our definition of minimum phase we see that, for $a(t)$ to be minimum phase, it is necessary that

$$a_0 > a_1. \tag{3.13}$$

We also see that, since $a(t)$ and $b(t)$ are inverses of each other, and $a(t)$ is causal and has finite energy, it follows that $b(t)$ must be minimum phase when $a(t)$ is minimum phase.

3.4. The Fourier Transform and Minimum Phase

Why is it called minimum phase? What is the phase? Consider the Fourier transform $A(f)$ of $a(t)$:

$$A(f) = \int_{-\infty}^{\infty} a(t)\, e^{-2\pi i f t} dt. \tag{3.14}$$

When $a(t)$ is the discrete two-sample time series, sampled at time intervals Δt,

$$a(t) = a_0, a_1, \tag{3.15}$$

we may write its discrete Fourier transform as (see Appendix)

$$A(f) = \sum_{n=0}^{1} a_n e^{-2\pi i f n \Delta t}. \tag{3.16}$$

The z-transform of $a(t)$ is given by equation (3.6), which we may write as

$$A(z) = \sum_{n=0}^{1} a_n z^n. \tag{3.17}$$

Comparing equations (3.16) and (3.17), we see that the z-transform and the Fourier transform are the same when

$$z = e^{-2\pi i f \Delta t}, \tag{3.18}$$

which is the unit circle in the complex plane.

Let us consider two possible cases

1. $a_0 = 1, a_1 = \frac{1}{2}$.

2. $a_0 = \frac{1}{2}, a_1 = 1$.

CASE 1. The Fourier transform of $a(t)$ is

$$A(f) = 1 + \frac{1}{2} e^{-2\pi i f \Delta t}. \tag{3.19}$$

In the complex plane this is the sum of two vectors: the first, the 1, is fixed and points from the origin to the point 1.0 along the real axis; the second, $\frac{1}{2} e^{-2\pi i f \Delta t}$, is a vector of length $\frac{1}{2}$ rotating clockwise as frequency f increases. The sum of these two vectors is a vector pointing from the origin to the point on the circle defined by the frequency f. As f changes, this vector changes in amplitude and phase according to

$$A(f) = |A(f)| e^{i\phi(f)}, \tag{3.20}$$

in which $|A(f)|$ is the *amplitude spectrum* and $\phi(f)$ is the *phase spectrum*.

The relationship between the complex plane (eq. 3.19) and the amplitude and the phase spectra (eq. 3.20) is shown in figure 3.1. The important thing to notice is the phase, which starts at a certain value (in this case 0.0), makes an excursion, passes through the zero frequency value again at $f = 1/2\Delta t$, makes an excursion of opposite polarity, and returns to the zero frequency value again at $f = 1/\Delta t$.

CASE 2. The Fourier transform of $a(t)$ is

$$A(z) = \frac{1}{2} + e^{-2\pi i f \Delta t}. \tag{3.21}$$

This consists of a fixed vector of amplitude $\frac{1}{2}$ and a rotating vector of amplitude 1. In the complex plane this circle actually encompasses the origin. As the frequency f increases, the second vector rotates, and the sum of the two vectors changes in amplitude in exactly the same way as for case 1. The

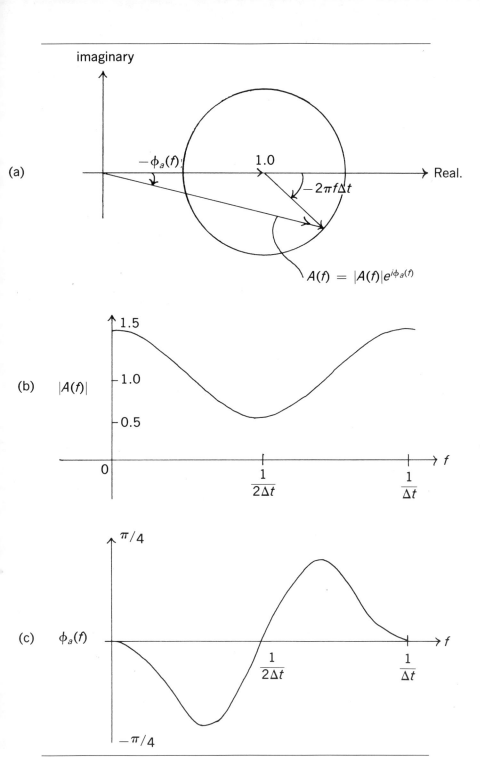

Figure 3.1. Relationship between (a) complex plane; (b) amplitude spectrum; and (c) phase spectrum of the two-point wavelet (1, ½).

phase is different, however: the vector $A(z)$ rotates completely, and at $f = 1/\Delta t$, the phase is greater by 2π than at $f = 0$. The z-transform and Fourier transform relationship for this case are shown in figure 3.2.

Thus, by turning our two-point wavelet 1, ½ back to front to produce the wavelet ½, 1, we have changed the phase spectrum and left the amplitude spectrum the same. The first wavelet is minimum-phase (as we have previously seen by definition), and we have shown pictorially that its phase is less than the phase of its time reverse.

Mathematically, we may describe the phase of our case 1 wavelet as

$$\tan(\phi_a(f)) = \frac{\text{½} \sin(-2\pi i f \Delta t)}{1 + \text{½} \cos(-2\pi i f \Delta t)}, \tag{3.22}$$

and the phase of our case 2 wavelet as

$$\tan(\phi_b(f)) = \frac{\sin(-2\pi i f \Delta t)}{\text{½} + \cos(-2\pi i f \Delta t)}. \tag{3.23}$$

We may rewrite equation (3.22) as

$$\tan(\phi_a(f)) = \frac{\sin(-2\pi i f \Delta t)}{2 + \cos(-2\pi i f \Delta t)}. \tag{3.24}$$

Comparing equations (3.23) and (3.24), we see immediately that the denominator on the right-hand side of equation (3.24) is larger than the denominator on the right-hand side of equation (3.23). Since the numerators are the same, it follows that

$$|\tan(\phi_b(f))| > |\tan(\phi_a(f))|, \tag{3.25}$$

and therefore that

$$|\phi_b(f)| > |\phi_a(f)|. \tag{3.26}$$

That is, the phase $|\phi_a(f)|$ is always less than $|\phi_b(f)|$ for all values of f. For this reason we call the case 1 wavelet the minimum-phase wavelet of the pair of two-length wavelets that share the same amplitude spectrum.

If we now have a longer wavelet $s(t)$, characterized by the samples

$$s_0, s_1, s_2, \ldots, s_n,$$

it will have a z-transform given by

$$S(z) = s_0 + s_1 z + s_2 z^2 + \ldots, s_n z^n. \tag{3.27}$$

We may factorize this z-transform into a number of roots, z_k, as follows:

$$S(z) = (-1)^n s_n \prod_{k=1}^{n} (z_k - z). \tag{3.28}$$

For the purposes of transformation into the frequency domain, each one of these factors $(z_k - z)$ can be written as a vector composed of a fixed part, z_k, and a rotating part

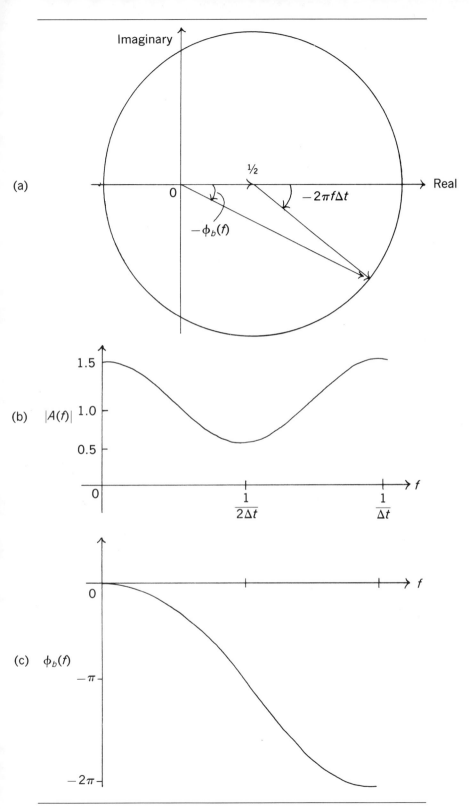

Figure 3.2. Relationship between (a) complex plane; (b) amplitude spectrum; and (c) phase spectrum of the two-point wavelet (½, 1).

$$z = e^{-2\pi i f \Delta t}.$$
(3.29)

Let the sum of these two vectors be the complex number c_k, where

$$z_k - z = c_k = |C_k(f)| e^{i\phi_k(f)}.$$
(3.30)

Substituting from equations (3.29) and (3.30) into equation (3.28), and recalling that multiplication of complex numbers consists of multiplication of their amplitudes and addition of their phases, we have

$$S(f) = (-1)^n s_n \prod_{k=1}^{n} |C_k(f)| \exp\left(\sum_{k=1}^{n} i\phi_k(f)\right).$$
(3.31)

Since the phases are additive, we see that the modulus of the phase spectrum of the wavelet $s(t)$ will be minimum for all frequencies f provided $|z_k| > 1$, for all k. That is, $s(t)$ is minimum-phase if all the roots of its z-transform $S(z)$ lie outside the unit circle.

We may check that this is consistent with our definition of minimum phase by considering the function $p(t)$, which we define to be the inverse of $s(t)$. Therefore, the z-transforms of $p(t)$ and $s(t)$ are related by the equation

$$P(z) = \frac{1}{S(z)}$$

$$= \frac{1}{(-1)^n s_n \prod_{k=1}^{n} (z_k - z)}.$$
(3.32)

We may expand the division in equation (3.32) as a series of partial fractions

$$P(z) = \sum_{k=1}^{n} \frac{Q_k}{(z_k - z)}.$$
(3.33)

If $p(t)$ is causal, $P(z)$ is expressed only in positive powers of z. If $|z_k| < 1$ for *any* of the n roots of $S(z)$, then the coefficients of the z-transform diverge with increasing powers of z. That is, $p(t)$ will not then have finite energy. The only way in which $p(t)$ can be causal *and* have finite energy is if *all* the roots z_k of $S(z)$ lie outside the unit circle, which is the result we wished to prove.

3.5. Power Spectrum, Autocorrelation Function, and Minimum Phase

We shall now look at the relationship between the power spectrum and the z-transform of a wavelet. Then, by looking at the roots of the z-transform corresponding to the power spectrum, we shall find a relationship between the number of wavelets that have the same power spectrum and the length of the spectrum. Finally, we shall identify the minimum-phase wavelet from the roots of the z-transform of the power spectrum.

If the Fourier transform $A(f)$ of a wavelet $a(t)$ is defined as in equation (3.14), the power spectrum is

$$\Phi_{aa}(f) = A(f) A^*(f) = |A(f)|^2, \tag{3.34}$$

where the asterisk superscript denotes complex conjugate. If the wavelet consists of discrete evenly spaced samples $a_0, a_1, \ldots, a_n$, and has the z-transform

$$A(z) = \sum_{k=0}^{n} a_k z^k, \tag{3.35}$$

we know that the z-transform and the discrete Fourier transform are identical when z is the unit circle defined by equation (3.18), which we reproduce here for convenience:

$$z = e^{-2\pi i f \Delta t}. \tag{3.18}$$

This unit circle is simply the locus of the points described by the end of a unit vector rotating clockwise. At a particular frequency f_1 the vector is frozen in a particular direction. The end of the vector is the complex number $e^{-2\pi i f_1 \Delta t}$. The *complex conjugate* of this number is $e^{2\pi i f_1 \Delta t}$, and from equation (3.18) we see that the equation

$$z^{-1} = e^{2\pi i f \Delta t} \tag{3.36}$$

describes the complex-conjugate unit circle. This is the same circle, but is described by the unit vector rotating *anticlockwise* at the same rate, as shown in figure 3.3. We see that when we deal with complex conjugates in the frequency domain, we must replace z by z^{-1} in the corresponding z-domain, if we are to be consistent.

Thus $A(f)$ is the Fourier transform of a wavelet $a(t)$ with z-transform $A(z)$, and $A^*(f)$ is the Fourier transform of a wavelet with z-transform $A^*(z)$. From equation (3.35) we can see that

$$A^*(z) = \sum_{k=0}^{n} a_k^* z^{-k}, \tag{3.37}$$

and we identify the corresponding wavelet as $a^*(-t)$. [We can substitute for $a^*(-t)$ into equation (3.14) to check that the Fourier transform is indeed $A^*(f)$.]

Turning now to the power spectrum defined by equation (3.34), we see that the corresponding z-transform is

$$\Phi_{aa}(z) = A(z) A^*(z), \tag{3.38}$$

which we may write as

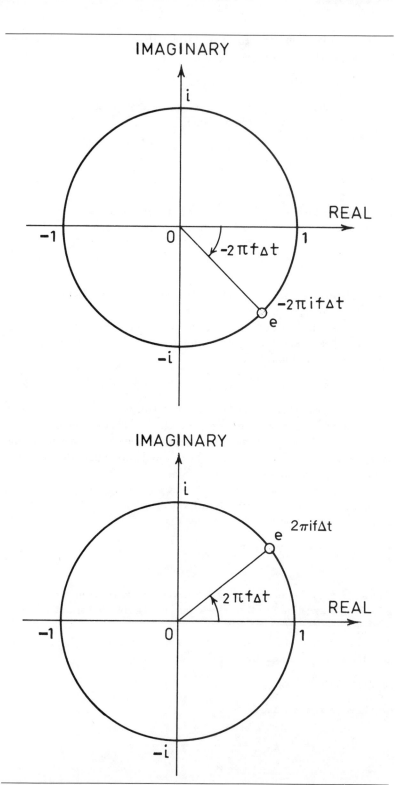

Figure 3.3. (a) The unit circle $z = e^{-2\pi i f\Delta t}$. (b) The unit circle $z^{-1} = e^{2\pi i f\Delta t}$.

$$\Phi_{aa}(z) = \sum_{\tau=-n}^{n} \phi_{aa}(\tau) \cdot z^{\tau}, \tag{3.39}$$

in which the $\phi_{aa}(\tau)$ are the autocorrelation coefficients.

It is now interesting to examine the roots of the polynomial $\Phi_{aa}(z)$. But $\Phi_{aa}(z)$ is not yet a polynomial because it contains negative as well as positive powers of z. This is because the autocorrelation $\phi_{aa}(\tau)$ of any wavelet $a(t)$ exists for both positive and negative values of τ. If we *delay* the autocorrelation by n time intervals, however, we can make it causal. Its z-transform will then contain terms in only positive powers of z. A delay of one time interval corresponds to multiplication by z in the z-domain. A delay of n time intervals corresponds to multiplication by z n times—that is, to multiplication by z^n. Thus we form the polynomial $G(z)$:

$$G(z) = z^n \Phi_{aa}(z) = A(z) z^n A^*(z). \tag{3.40}$$

We can factorize the z-transform $A(z)$ into its n roots:

$$A(z) = (-1)^n a_n \prod_{k=1}^{n} (z_k - z). \tag{3.41}$$

Similarly, we can factorize the z-transform $z^n A^*(z)$ into its n roots:

$$z^n A^*(z) = z^n \cdot (-1)^n a_n^* \prod_{k=1}^{n} (z_k^* - z^{-1})$$

$$= a_n^* \prod_{k=1}^{n} (1 - z_k^* z). \tag{3.42}$$

Combining equations (3.40), (3.41), and (3.42), we have

$$G(z) = g_{2n} \prod_{k=1}^{n} (z_k - z)(1 - z_k^* z), \tag{3.43}$$

where

$$g_{2n} = (-1)^n a_n \cdot a_n^*. \tag{3.44}$$

The roots z_k and $1/z_k^*$ are arranged symmetrically with respect to the unit circle, as shown in figure 3.4, where we have taken $|z_k| > 1$. It is clear that the $2n$ roots occur in n pairs, each of which contains one root inside the unit circle and one outside. It is possible that $|z_k| = 1$, in which case both roots lie on the unit circle. This is a limiting case that poses theoretical problems for determination of the phase spectrum. Within the accuracy of real data, however, it is always possible to consider that a root on the unit circle is either just inside or just outside the circle, whichever is more convenient. If the autocorrelation is real, the roots occur in complex conjugate pairs, as shown in figure 3.5.

Let us now suppose that we are confronted with the problem of trying to find the wavelet $a(t)$, given its power spectrum $\Phi_{aa}(f)$. How do we do it? First

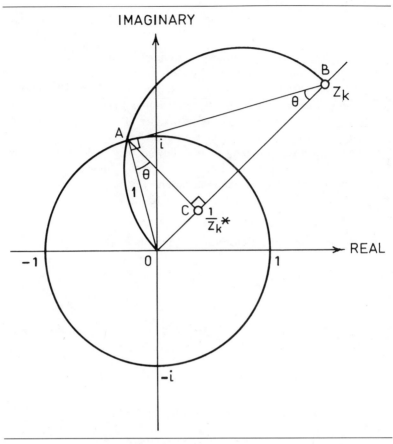

Figure 3.4. *Showing symmetry of the roots* z_k *and* $1/z_k{}^*$ *with respect to the unit circle. These roots are at B and C respectively. OCB is a straight line and diameter of the circle which intersects the unit circle at A. C is located at the foot of the perpendicular from A to OB. Comparing the two right similar triangles OAB and OCA we see* $\sin\theta = OC/OA = OA/OB$. *Therefore,* $OC \cdot OB = OA^2 = 1$. *Therefore,* $OC = 1/OB = 1/|z_k|$.

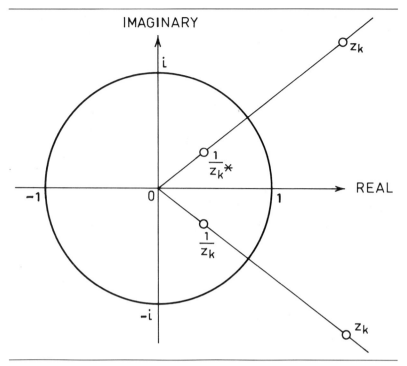

Figure 3.5. Showing symmetry of roots of autocorrelation function.

we could find the Fourier transform of $\Phi_{aa}(f)$, which is the autocorrelation function $\phi_{aa}(\tau)$. We can then write down the z-transform $\Phi_{aa}(z)$ from equation (3.39) and multiply it by z^n to form the polynomial $G(z)$, as in equation (3.40). Then we could factorize the polynomial to find the $2n$ roots. We then simply select the n roots that correspond to the z-transform $A(z)$ of our wavelet $a(t)$.

We must select one root from each of the n pairs of roots. Which root do we select, the one inside or the one outside the unit circle? We have to choose between these two alternatives for every one of the n pairs. That is, there are 2^n possible different combinations of roots we can select. Each one of these combinations of roots forms a different z-transform, corresponding to a different wavelet $a(t)$. All these different wavelets have the same power spectrum $\Phi_{aa}(f)$ and the same autocorrelation $\phi_{aa}(\tau)$.

We see that we cannot find the wavelet uniquely when we know only the power spectrum or the autocorrelation function. There are 2^n possible different wavelets. To reduce the number of possible choices, we need more information. Suppose, for example, we know that the wavelet is minimum-phase. The selection of the roots is then easy.

We have proved in section 3.3 that a minimum-phase wavelet has all the roots of its z-transform outside the unit circle. Since the z-transform $G(z)$ has n roots outside the unit circle and n roots inside, there is a unique combination of n roots that forms the z-transform of our minimum-phase wavelet.

3.6. Robinson's Energy-Delay Theorem and Minimum Phase

Enders Robinson (1962) showed that of all the causal wavelets that have the same power spectrum, the minimum-phase wavelet has its energy concentrated as close to the beginning as possible. We can express this a little more mathematically, as follows.

If $a_{min}(t)$ is a minimum-phase wavelet that has the same power spectrum as a certain suite of wavelets, and $a(t)$ is a non-minimum-phase wavelet in this suite, then

$$E_{a_{min}}(\tau) \geq E_a(\tau), \qquad \text{for all } \tau, \tag{3.45}$$

where

$$E_{a_{min}}(\tau) = \int_0^\tau a^2_{min}(t)\, dt, \tag{3.46}$$

and

$$E_a(\tau) = \int_0^T a^2(t) \, dt. \tag{3.47}$$

Another way of expressing the same thing is: the minimum-phase wavelet has the fastest energy buildup, or greatest partial energy of all the 2^n sequences in the suite with the same power spectrum. A proof of this for discrete data is in Claerbout (1976, p. 52).

3.7. Minimum Phase and the Wiener Inverse Filter

There is an intimate relationship between the Wiener filter and the concept of minimum phase, and it pervades everything that is normally done in seismic data processing. In this section we shall show that the Wiener inverse filter is minimum-phase. [The exact inverse is defined by equation (3.3).]

In chapter 1 we described the Wiener filter f_t, which, when convolved with an input x_t, will give an output y_t, which is a least-squares approximation to some desired output z_t. The equations we must solve to find the coefficients of the filter are known as the *normal equations* and are written as equations (1.20).

Suppose we have a sequence $(a_0, a_1, \ldots, a_n)$ and wish to find its least-squares inverse $(f_0, f_1, \ldots, f_n)$. We find it by solving the normal equations for a_t as input and

$$\begin{aligned} z_t &= 1, & t &= 0 \\ &= 0, & t &\neq 0 \end{aligned} \tag{3.48}$$

as desired output. The actual output y_t of the filter is given by the convolution

$$y_t = \sum_{s=0}^{n} f_s a_{t-s}. \tag{3.49}$$

Neither f_t nor a_t has any values before $t = 0$; therefore,

$$y_0 = f_0 \, a_0. \tag{3.50}$$

The energy in the sequence y_t is equal to

$$\sum_{t=0}^{2n} y_t^2 = \phi_{yy}(0). \tag{3.51}$$

The error energy is given by the expression

$$I = \sum_{t=0}^{2n} (z_t - y_t)^2. \tag{3.52}$$

Using the definitions of z_t and y_t given earlier, we find

$$I = (1 - y_0)^2 + \sum_{t=1}^{2n} y_t^2$$

$$= 1 - 2y_0 + y_0^2 + \sum_{t=1}^{2n} y_t^2$$

$$= 1 - 2y_0 + \phi_{yy}(0). \qquad (3.53)$$

From equation (3.53) and Robinson's energy-delay theorem, we shall show, following Robinson (1967, p. 171–173) that

1. The error energy I is least when a_t is minimum-delay.
2. f_t is minimum-delay.

The z-transform of the autocorrelation $\phi_{yy}(\tau)$ of the output y_t is

$$\Phi_{yy}(z) = Y(z)\,Y^*(z)$$
$$= F(z)\,A(z) \cdot F^*(z)\,A^*(z)$$
$$= \Phi_{ff}(z)\,\Phi_{aa}(z). \qquad (3.54)$$

Therefore

$$\phi_{yy}(\tau) = \phi_{ff}(\tau) * \phi_{aa}(\tau). \qquad (3.55)$$

Thus the suite of sequences $\{a_t\}$ that has the same autocorrelation function $\phi_{aa}(\tau)$ will form a suite of outputs $\{y_t\}$ that all have the same autocorrelation $\phi_{yy}(\tau)$. Therefore, $\phi_{yy}(0)$ is a constant for this suite of input sequences $\{a_t\}$.

We may rewrite equation (3.53) as

$$I = [1 + \phi_{yy}(0)] - 2y_0. \qquad (3.56)$$

Now $\phi_{yy}(0)$ is positive, being a sum of squares; and y_0 must also be positive; if it were not, it would be possible to reduce I by changing the algebraic sign of the coefficients of f_t. Since, however, by definition, the f_t have been obtained from the condition that I is a minimum (see eq. 1.14), f_0 and a_0 must have the same sign.

If we now apply Robinson's energy-delay theorem to the suite $\{a_t\}$, we can deduce that the minimum-phase wavelet has its energy concentrated as close to the beginning $t = 0$ as possible; in particular, the energy in the first coefficient a_0 is bigger for the minimum-phase wavelet than it is for any other wavelet in the suite. That is, a_0 is a maximum for the minimum-phase wavelet.

It follows that $y_0 = f_0 a_0$ will be a maximum if a_t is minimum-phase, and I must be a minimum if a_t is minimum-phase. This is the first result we wished to show.

It is similarly true that of all the sequences f_t in the suite $\{f_t\}$ sharing the same autocorrelation $\phi_{ff}(\tau)$, the minimum-phase sequence will have its first coefficient f_0 a maximum. Now the output y_t has the same autocorrelation ϕ_{yy} (τ) for all filters in the suite $\{f_t\}$ having the autocorrelation ϕ_{ff} (τ). ϕ_{yy} (0) is therefore a constant for this suite. If f_t were *not* minimum-phase, it would be possible to reduce I in equation (3.56) by choosing $f_0 = f_{0\ min}$. Since, however, I is by definition a minimum, f_0 must be a maximum. Therefore, by Robinson's energy-delay theorem, f_t is minimum-phase. We note that this is true independent of the phase of a_t. This is the second result we wished to prove.

These two results are interesting. The shape of the filter f_t depends only on the autocorrelation function ϕ_{aa} (τ), and not on the phase of the actual input a_t. The filter f_t is fixed and is minimum-phase. It does a better job of inverting a_t when a_t is minimum-phase than when a_t is not minimum-phase. Therefore, the error energy I depends on the phase of a_t and is least when a_t is minimum-phase.

The minimum-phase property of the least-squares inverse filter f_t can be used to find the minimum-phase wavelet in the suite $\{a_t\}$, given any other wavelet in the suite. Let a_t be one wavelet in the suite $\{a_t\}$ having a given autocorrelation function ϕ_{aa} (τ). We find the least-squares inverse f_t of a_t as discussed earlier. It happens to be minimum-phase, as shown above. If we now find the autocorrelation of f_t, we can find its least-squares inverse from the normal equations. Let us call this filter $\hat{a}_{t\ min}$. It will be minimum-phase and will be the least-squares inverse of f_t, which itself is the least-squares inverse of a_t. Thus $\hat{a}_{t\ min}$ is a least-squares approximation to the minimum-phase wavelet of the suite $\{a_t\}$.

In the next chapter we shall see that the relationship between minimum phase and Wiener filtering extends to prediction and prediction-error filtering.

4. PREDICTION AND PREDICTION-ERROR
FILTERING

4.1. Introduction

In seismic data processing the most commonly used initial data processing sequence is as follows:

1. Demultiplex.
2. Spherical-divergence correction.
3. Predictive deconvolution (gap deconvolution).

After demultiplexing, to arrange the data in trace-sequential format, a *spherical-divergence correction* is applied to each trace of the record. As discussed in chapter 2, the spherical-divergence correction is applied to force the data to appear as if they had been obtained with a plane-wave source. As further discussed in chapter 2, this correction does not really achieve that objective. Nevertheless, it is still applied.

Despite the obvious objections that can be raised against the spherical-divergence correction on the grounds of its geometrical nonsense and its uneven distortion of the source wavelet, there is a very powerful argument in its favor: it makes the data more nearly stationary. The assumption of stationarity is crucial to most statistical deconvolution techniques. Moreover, since seismic reflection data are notoriously nonstationary, they must be *forced* to be close to stationary—so that the stationarity assumption can be applied.

In approaching the problem of predictive deconvolution, we bear in mind that the so-called spherical-divergence correction is a data-massaging device that distorts the data. This distortion is tolerated as the price that must be paid when stationarity is so highly valued.

An example of a demultiplexed record before and after spherical-divergence correction is shown in figure 4.1.

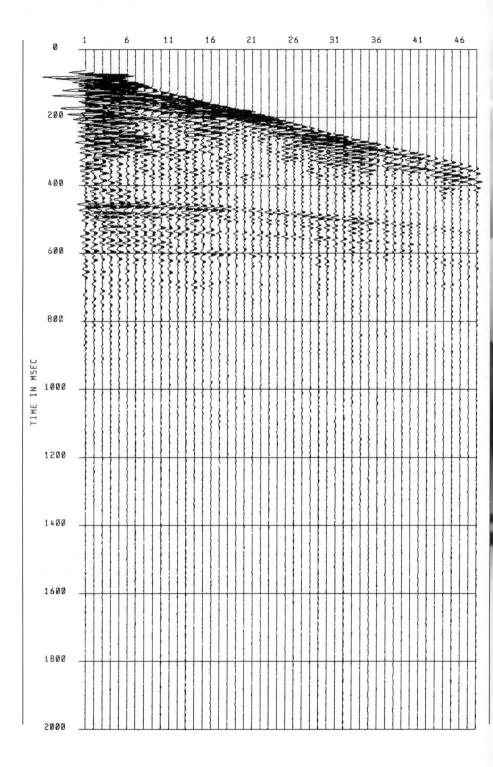

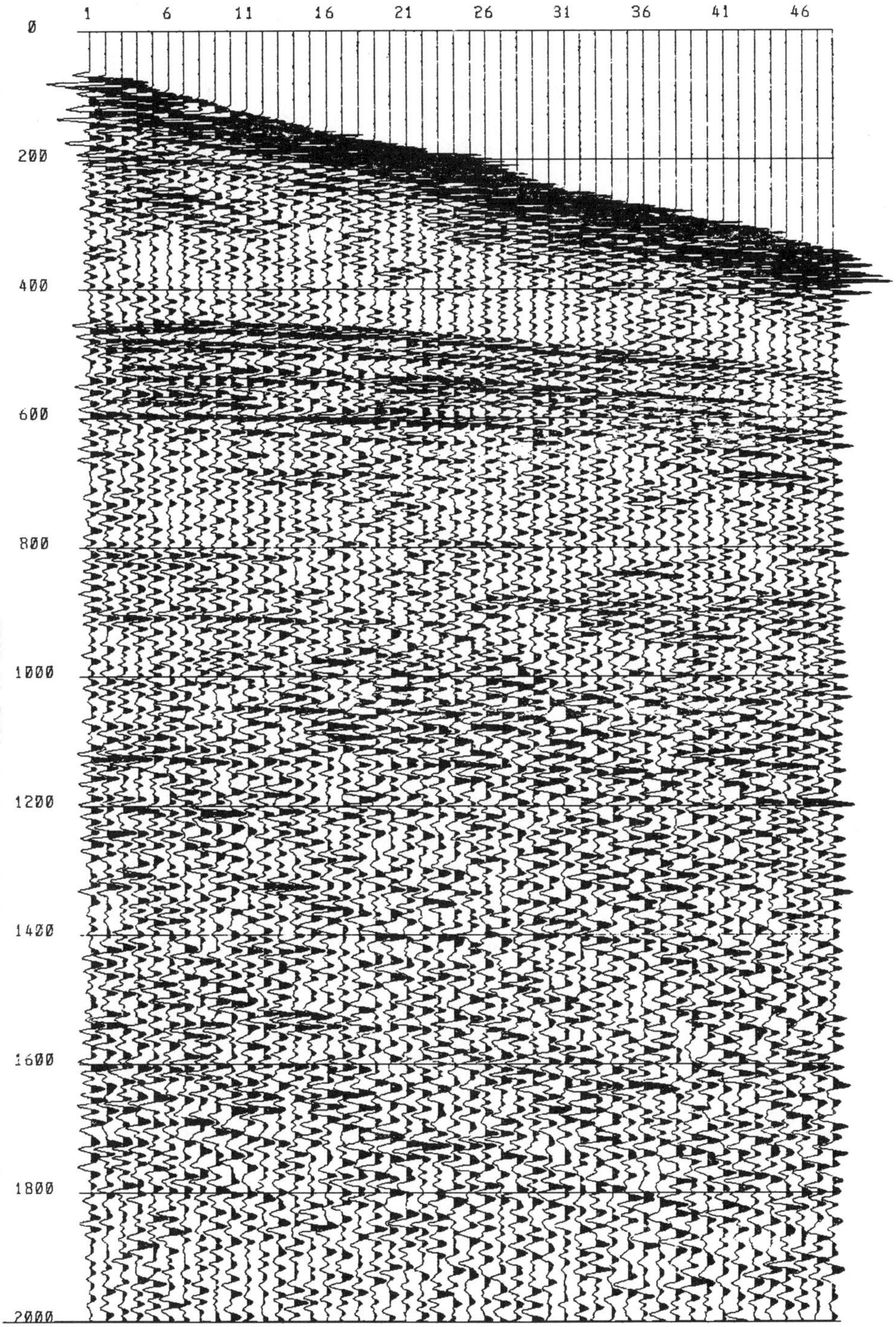

Figure 4.1(b). The same record as 4.1(b) after spherical divergence correction and trace equalization.

4.2. The Prediction Filter

The problems of prediction and deconvolution are very closely related, as pointed out by Peacock and Treitel (1969) in their excellent paper "Predictive Deconvolution: Theory and Practice." In what follows, we retrace the steps of Peacock and Treitel showing, in parallel with their mathematical development, that the prediction error filter can be derived from the prediction filter by the use of diagrams and no mathematics.

If p_t is a *prediction filter* $p_0, p_1, \ldots, p_n$, with prediction distance α, its output y_t is an estimate of the future value of the input x_t at some time $t + \alpha$. We thus write

$$y_t = \sum_{s=0}^{n} p_s x_{t-s} = \hat{x}_{t+\alpha},\tag{4.1}$$

where $\hat{x}_{t+\alpha}$ is an estimate of $x_{t+\alpha}$. If we wish to obtain the least-squares (Wiener) prediction filter, we simply require the desired output z_t to equal $x_{t+\alpha}$, the time-advanced version of the input x_t. The right-hand side of the normal equations (section 1.2) becomes:

$$\phi_{zx}(\tau) = \sum_{t=\tau}^{\infty} x_{t+\alpha} x_{t-\tau}$$

$$= \sum_{s=\tau+\alpha}^{\infty} x_s\, x_{s-(\tau+\alpha)}$$

$$= \phi_{xx}(\tau+\alpha).\tag{4.2}$$

The normal equations thus become:

$$\sum_{s=0}^{n} p_s\, \phi_{xx}(s - \tau) = \phi_{xx}(\tau + \alpha), \qquad \tau = 0, 1, \ldots, n.\tag{4.3}$$

The solution to these equations yields the Wiener prediction filter of prediction distance α. We see that this filter depends only on the autocorrelation of the input, and not on the phase of the input. This in itself is interesting. We know from chapter 3 that there are many different sequences that can share the same autocorrelation function. Yet, despite all these differences, there is a unique least-squares filter $p_0, p_1, \ldots, p_n$ that will give us our best estimate of the future value of x_t at time $t + \alpha$. This is our best estimate *whatever* the phase of the input x_t. In figure 4.2 we show the convolution of the input x_t with the filter p_t.

If we store our estimated value

$$\hat{x}_{t + \alpha} = y_t,$$

made at time t, and wait until time $t + \alpha$, we can find the error $\epsilon_{t+\alpha}$ that we have made in our prediction. The error is defined as

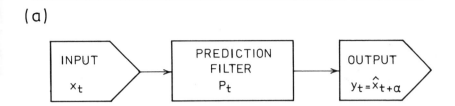

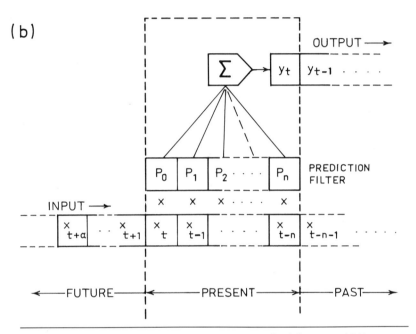

Figure 4.2. Linear prediction of x_{t+2} from samples of x_t up to and including time t. (a) Linear system description. (b) Sampled signal description.

$$\epsilon_{t+a} = x_{t+a} - \hat{x}_{t+a}. \tag{4.4}$$

The error series represents the nonpredictable part of x_{t+a}. That is, the estimate $\hat{x}_{t+a}$ is accurate only to a certain extent. It is derived from $n + 1$ filter coefficients and knowledge of x from time $t - n$ to time t, as shown in figure 4.2. Between time t and time $t + a$, the process that is generating x_t may be, to some extent, unpredictable.

If, for example, x_t is a random, stationary, white-noise sequence with zero mean (see Appendix), then every sample is independent of every other sample. Knowledge of x from times $t - n$ to t does not help in estimating the value of x at time $t + a$. This becomes apparent when we consider the autocorrelation $\phi_{xx}(\tau)$ of x_t. If x_t is truly white, random, and stationary with zero mean, then

$$\begin{aligned} \phi_{xx}(\tau) &= 0, \quad \tau \neq 0 \\ &= \sigma^2, \quad \tau = 0, \end{aligned} \tag{4.5}$$

where σ^2 is the variance (see Appendix) of the sequence x_t. This situation is illustrated in figure 4.3. In the normal equations (4.3) all the terms on the right-hand side are zero for any nonzero value of a. Therefore, all the filter coefficients are zero. Further, the best estimate of the value of x at time $t + a$ is zero. That is, the best estimate is the mean because every sample is totally independent of every other sample.

If, however, x_t is random and stationary but not white, then its autocorrelation function will not be a spike at zero lag. It will have some shape, with values at lags τ that are nonzero. The prediction filter coefficients will not then necessarily be zero. If the autocorrelation function is a narrow function confined to the region

$$- T \leq \tau \leq T, \tag{4.6}$$

such that

$$\phi_{xx}(\tau) = 0, \quad \tau > |T|, \tag{4.7}$$

as illustrated in figure 4.4, then any prediction filter can be effective only up to prediction distances a that are not greater than T. At greater distances the right-hand side of the normal equations (4.3) is always zero, and the best estimate becomes zero, as before.

We thus see that the accuracy of our prediction depends on: (1) the nature of the process that is generating the series x_t; (2) the number $n + 1$ of filter coefficients at our disposal; and (3) the prediction distance a. The error in our prediction is a measure of the *un*predictable part of the process.

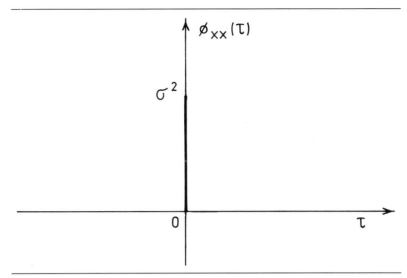

Figure 4.3. *Autocorrelation of a white random stationary noise sequence* x_t *with zero mean and variance* σ^2.

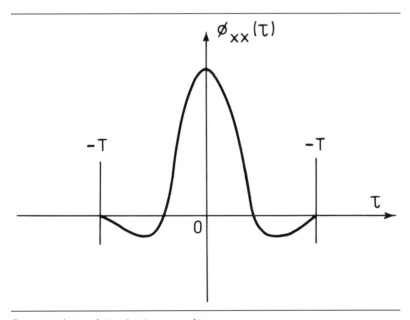

Figure 4.4. *Autocorrelation of stationary nonwhite sequence* x_t.

4.3. The "Prediction-Error" Filter

We can replace $\hat{x}_{t+\alpha}$ in equation (4.4) by its definition in equation (4.1):

$$\epsilon_{t+\alpha} = x_{t+\alpha} - y_t$$

$$= x_{t+\alpha} - \sum_{s=0}^{n} p_s x_{t-s}. \tag{4.8}$$

In order to do the subtraction in equation (4.8), we have to store the value y_t, created at time t, and wait until time $t + \alpha$. At time $t + \alpha$ we find out what the value of x is and can subtract our estimate y_t from it. This operation is illustrated in figure 4.5.

We may take the z-transform of equation (4.8) as follows:

$$z^{-\alpha} E(z) = z^{-\alpha} X(z) - P(z) \cdot X(z), \tag{4.9}$$

which we may write as

$$E(z) = X(z) [1 - z^{\alpha} P(z)]. \tag{4.10}$$

We could have arrived at the same result by saying that the z-transform of the error series is $E(z)$; the z-transform of x_t is $X(z)$; the z-transform of y_t is $Y(z) = P(z).X(z)$; but we must delay this by α samples before the subtraction can be made. Thus,

$$E(z) = X(z) - z^{\alpha} P(z) X(z)$$
$$= X(z) [1 - z^{\alpha} P(z)]. \tag{4.11}$$

We see immediately that the factor $[1 - z^{\alpha} P(z)]$ is the z-transform of the filter

$$1, 0, 0, \ldots, 0, -p_0, -p_1, \ldots, -p_n,$$

$$\alpha - 1 \text{ zeros}$$

and the error series ϵ_t is simply the convolution of the series x_t with this filter. This is illustrated in figure 4.6. Comparing figure 4.6 with figure 4.5, we can see that the two situations produce the error series $\epsilon_{t+\alpha}$ either by convolving x_t with p_t to form y_t and then subtracting y_t from $x_{t+\alpha}$, or by convolving the series x_t with the filter

$$1, 0, 0, \ldots, 0, -p_0, -p_1, \ldots, -p_n$$

$$\alpha - 1 \text{ zeros}$$

directly.

This filter is called the prediction-error filter, which is something of a misnomer. This so-called prediction-error filter does *not* predict the error (if it did, we would be able to make the prediction exact). We can see from the

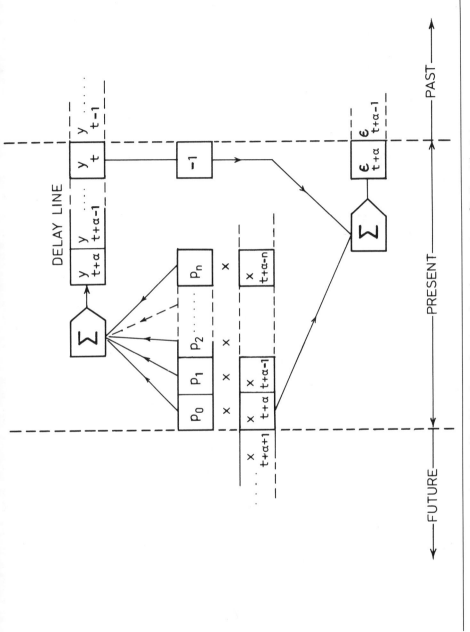

Figure 4.5. Prediction of future values of x_t, followed by a later subtraction to form the error series ϵ_t. Situation at time $t + \alpha$.

65

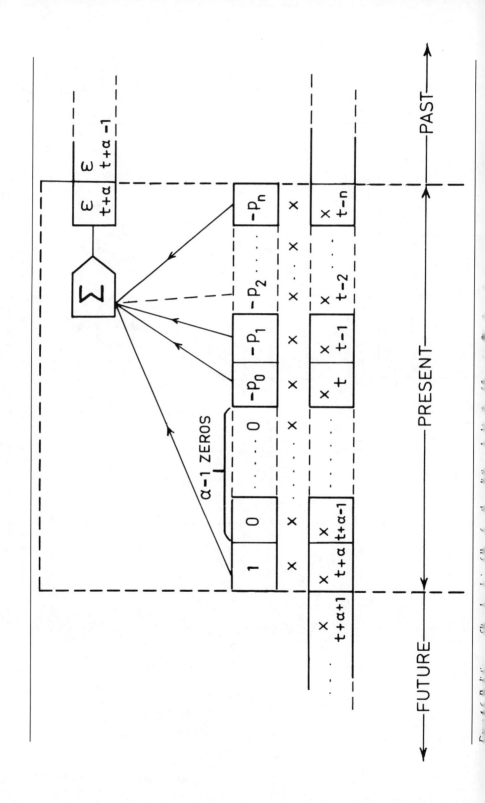

preceding argument that the error can be calculated only when the value of x at time $t + \alpha$ is known.

4.4. Relationship between Prediction-Error
Filtering and Inverse Filtering

This section follows the arguments of Peacock and Treitel (1969). It will now be shown that the Wiener inverse filter, which ideally transforms a *real* signal of known autocorrelation to an impulse at zero delay, is equivalent to the prediction-error filter corresponding to the one-step-ahead prediction filter (for which $\alpha = 1$). The normal equations for the $n + 1 -$ length prediction filter p_t, which predicts the next sample $(\alpha + 1)$, are:

$$\phi_{xx}(0)\, p_0 + \quad \phi_{xx}(1)\, p_1 + \ldots + \phi_{xx}(n)\, p_n = \phi_{xx}(1).$$

$$\phi_{xx}(-1)\, p_0 + \phi_{xx}(0)\, p_1 + \ldots + \phi_{xx}(n - 1)\, p_n = \phi_{xx}(2).$$

$$\phi_{xx}(-n)p_0 + \phi_{xx}(-n + 1)p_1 + \ldots + \phi_{xx}(0)p_n = \phi_{xx}(n + 1). \tag{4.12}$$

We recall from section 1.2 that the autocorrelation of a real signal is symmetrical. That is,

$$\phi_{xx}(\tau) = \phi_{xx}(-\tau) \tag{4.13}$$

(see section 1.2). Equations (4.12) can therefore be written as:

$$\phi_{xx}(0)\, p_0 + \quad \phi_{xx}(1)\, p_1 + \ldots + \phi_{xx}(n)\, p_n = \phi_{xx}(1).$$

$$\phi_{xx}(1)\, p_0 + \phi_{xx}(0)\, p_1 + \ldots + \phi_{xx}(n - 1)\, p_n = \phi_{xx}(2).$$

$$\phi_{xx}(n)\, p_0 + \phi_{xx}\, (n - 1)\, p_1 + \ldots + \phi_{xx}(0)\, p_n = \phi_{xx}(n + 1). \tag{4.14}$$

We now modify these equations by first subtracting the coefficient $\phi_{xx}(i)$ from both sides of the ith row of each equation such that the right-hand side vanishes:

$$-\phi_{xx}(1) \quad + \phi_{xx}(0)p_0 + \phi_{xx}(1)p_1 \quad + \ldots + \phi_{xx}(n)p_n \quad = \phi_{xx}(1) \quad -\phi_{xx}(1).$$

$$-\phi_{xx}(2) \quad + \phi_{xx}(1)p_0 + \phi_{xx}(0)p_1 \quad + \ldots + \phi_{xx}(n-1)p_n = \phi_{xx}(2) \quad -\phi_{xx}(2).$$

$$\vdots \qquad \qquad \vdots \qquad \quad \vdots \qquad \qquad \vdots \qquad \qquad \vdots$$

$$-\phi_{xx}(n+1) + \phi_{xx}(n)p_0 + \phi_{xx}(n-1)p_1 + \ldots + \phi_{xx}(0)p_n \quad = \phi_{xx}(n+1) -\phi_{xx}(n+1).$$

$$\tag{4.15}$$

We then augment the system by adding another equation at the top:

$$-\phi_{xx}(0) \quad + \phi_{xx}(1)p_0 + \phi_{xx}(2)p_1 \quad + \ldots + \phi_{xx}(n+1)p_n = -\beta$$

$$-\phi_{xx}(1) \quad + \phi_{xx}(0)p_0 + \phi_{xx}(1)p_1 \quad + \ldots + \phi_{xx}(n)p_n \quad = \phi_{xx}(1) \quad -\phi_{xx}(1).$$

$$-\phi_{xx}(2) \quad + \phi_{xx}(1)p_0 + \phi_{xx}(0)p_1 \quad + \ldots + \phi_{xx}(n-1)p_n = \phi_{xx}(2) \quad -\phi_{xx}(2).$$

$$\vdots \qquad\qquad \vdots \qquad \vdots \qquad\qquad \vdots \qquad\qquad \vdots \qquad\qquad \vdots$$

$$-\phi_{xx}(n+1) + \phi_{xx}(n)p_0 + \phi_{xx}(n-1)p_1 + \ldots + \phi_{xx}(0)p_n \quad = \phi_{xx}(n+1) - \phi_{xx}(n+1).$$

$$(4.16)$$

This system may be written

$$\phi_{xx}(0) - \quad \phi_{xx}(1)p_0 - \phi_{xx}(2)p_1 - \ldots - \phi_{xx}(n+1)p_n = \beta.$$

$$\phi_{xx}(1) - \quad \phi_{xx}(0) - \phi_{xx}(1)p_1 - \ldots - \quad \phi_{xx}(n)p_n = 0.$$

$$\vdots \qquad\qquad \vdots \qquad \vdots \qquad\qquad \vdots$$

$$\phi_{xx}(n+1) - \phi_{xx}(n)p_0 - \phi_{xx}(n-1)p_1 - \ldots - \quad \phi_{xx}(0)p_n = 0. \quad (4.17)$$

for which the associated matrix equation is:

$$\begin{bmatrix} \phi_{xx}(0) & \phi_{xx}(1) & \phi_{xx}(n+1) \\ \phi_{xx}(1) & \phi_{xx}(0) & \phi_{xx}(n) \\ \vdots & & \\ \phi_{xx}(n+1) & \phi_{xx}(n) & \phi_{xx}(0) \end{bmatrix} \begin{bmatrix} 1 \\ -p_0 \\ \vdots \\ -p_n \end{bmatrix} = \begin{bmatrix} \beta \\ 0 \\ \vdots \\ 0 \end{bmatrix}. \quad (4.18)$$

We see that the Wiener filter 1, $-p_0$, $-p_1$, $-p_2$, $\ldots$, $-p_n$, which is the solution of equations (4.18), can be identified immediately as the prediction-error filter corresponding to the prediction filter of equations (4.17). Equations (4.18) can be rewritten in the form:

$$\begin{bmatrix} \phi_{xx}(0) & \phi_{xx}(1) \ldots \phi_{xx}(n+1) \\ \phi_{xx}(1 & \phi_{xx}(0) \ldots \phi_{xx}(n) \\ \vdots & \\ \phi_{xx}(n+1) & \phi_{xx}(n) \ldots \phi_{xx}(0) \end{bmatrix} \begin{bmatrix} b_0 \\ b_1 \\ \vdots \\ b_{n+1} \end{bmatrix} = \begin{bmatrix} \beta \\ 0 \\ \vdots \\ 0 \end{bmatrix}. \quad (4.19)$$

where

$$b_0 = 1.$$

$$b_i = -p_{i-1}, \quad i = 1, 2, \ldots, n+1.$$

$$\beta = \sum_{i=0}^{n+1} b_i \, \phi_{xx}(i). \quad (4.20)$$

We now note that equations (4.19) are very similar to the equations (1.31) for the spiking filter. In fact, if x is real, as we have assumed, they are identical, apart from the factor on the right-hand side and the number of equations. Since we could just as well have started with a prediction filter of arbitrary length n, instead of length $n + 1$, this difference is trivial. The factor β affects only the amplitude of the filter, not its shape. [Equations (1.30) and (1.31) show how this amplitude scaling works.]

In section 3.7 we showed that the Wiener inverse filter, or spiking filter, derived from the normal equations, is minimum-phase. Since the one-step-ahead prediction-error filter is identical with the Wiener inverse filter, we see that it too must be minimum-phase. It must follow that the Wiener prediction filter has a special property. We may find what is special about it by considering the errors in prediction. (In the industry the one-step-ahead prediction-error filter is sometimes called a "spiking filter." This is confusing since it is based on the autocorrelation of the received data, which may or may not be the same as the autocorrelation of the wavelet. In this book a spiking filter turns a known wavelet into a spike.)

4.5. The Errors in Prediction

Consider the stationary sequence x_t, which is the result of the convolution of a stationary white-noise sequence q_t with the linear filter a_t, as illustrated in figure 4.7. Thus

$$x_t = a_t * q_t. \tag{4.21}$$

Let the white-noise sequence have zero mean and variance σ_q^2:

$$E\{q_t\} = 0.$$

$$\begin{aligned} E\{q_t \cdot q_s\} &= \phi_{qq}(t-s) \\ &= \sigma_q^2, \; s=t \\ &= 0, \; s \neq t. \end{aligned} \tag{4.22}$$

The expectation function E is basically the same as the autocorrelation function (See section 1.2 and Appendix). Thus we can say that q_t has an autocorrelation

$$\begin{aligned} \phi_{qq}(\tau) &= \sigma_q^2, \quad \tau = 0 \\ &= 0, \quad \tau \neq 0. \end{aligned} \tag{4.23}$$

In equation (3.55) of the previous chapter we showed that the autocorrelation of the filter output is equal to the autocorrelation of the filter convolved with the autocorrelation of the input. Thus:

$$\begin{aligned} \phi_{xx}(\tau) &= \phi_{qq}(\tau) * \phi_{aa}(\tau) \\ &= \sigma_q^2 \, \delta(\tau) * \phi_{aa}(\tau) \\ &= \sigma_q^2 \, \phi_{aa}(\tau). \end{aligned} \tag{4.24}$$

Now consider the problem of predicting x_{t+1} from the values of x_t up to and including time t. The equations we must solve are equations (4.3), with $\alpha = 1$. The corresponding one-step-ahead prediction-error filter $(1, -p_0, -p_1, \ldots, -p_n)$ can be found directly from equation (4.19). If we now substitute for $\phi_{xx}(\tau)$ from equation (4.24) into equations (4.19), we find:

$$
\begin{bmatrix}
\phi_{aa}(0) & \phi_{aa}(1) \ldots \phi_{aa}(n+1) \\
\phi_{aa}(1) & \phi_{aa}(0) \ldots \phi_{aa}(n) \\
\vdots & \\
\phi_{aa}(n+1) & \phi_{aa}(n) \ldots \phi_{aa}(0)
\end{bmatrix}
\begin{bmatrix}
b_0 \\
b_1 \\
\vdots \\
b_{n+1}
\end{bmatrix}
=
\begin{bmatrix}
\beta/\sigma_q^2 \\
0 \\
\vdots \\
0
\end{bmatrix}. \quad (4.25)
$$

where b_t is the prediction error filter and β is the scale factor defined by

$$
\beta = \sum_{i=0}^{n+1} b_i\, \phi_{aa}(i),
$$

which we could rewrite as

$$
\frac{\beta}{\sigma_q^2} = \sum_{i=0}^{n+1} b_i\, \phi_{aa}(i) = \mu. \quad (4.26)
$$

We immediately recognize the prediction filter b_t as the Wiener approximation to the inverse of a_t. That is, the convolution of b_t with a_t is approximately a spike at time $t = 0$. In our present case the input to a_t is the white-noise sequence q_t, and the output is the sequence x_t. The result of filtering x_t with the prediction-error operator b_t is approximately to recover the white-noise sequence q_t. In other words, if our prediction-error filter is the *exact* inverse of a_t (which is the best we can hope to do) the error series ϵ_t will exactly equal the white-noise series q_t. Therefore, q_t represents the nonpredictable part of x_t, as illustrated in figure 4.8.

There are three limitations to the accuracy of our prediction:

1. In practice, x_t is always a series of finite length, with errors. Estimates of the autocorrelation function $\phi_{xx}(\tau)$ are therefore always inexact, even if x_t is a stationary linear process, as we have supposed here.
2. The prediction-error filter $(b_0, b_1, \ldots, b_{n+1})$ is a realizable sequence with no anticipatory component $(b_{-1}, b_{-2}, \ldots)$ and can be the exact inverse of a_t only if a_t is minimum-phase. Even if a_t *is* minimum-phase, b_t is unlikely to be an exact inverse because it is of finite length—restricted by the length of the autocorrelation $\phi_{xx}(\tau)$.
3. The linear filter a_t has an exact causal inverse only if it is minimum-phase (see chapter 3); but because there are many filters that have the same autocorrelation as a_t, we do not know a priori that a_t is minimum-phase.

Let us now consider two situations: (1) where b_t is the exact inverse of $a_{t\ \min}$, the minimum-phase wavelet with autocorrelation $\phi_{aa}(\tau)$; and (2)

Figure 4.7. A stationary sequence x_t can be regarded as the convolution of a stationary white noise sequence q_t with a linear smoothing filter a_t.

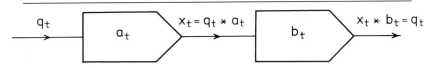

Figure 4.8. Prediction-error filtering of the stationary sequence x_t. The stationary white random noise input is the unpredictable part of x_t. The prediction-error filter b_t can return q_t exactly if a_t is minimum-phase and if b_t is the exact inverse of a_t.

where b_t is an approximation to the inverse of $a_{t\,min}$. In each situation we shall consider the case of $a_t = a_{t\,min}$ and $a_t \neq a_{t\,min}$.

1. b_t is the exact inverse of $a_{t\,min}$.
   a. If $a_t = a_{t\,min}$, we have

$$a_t * b_t = a_{t\,min} * b_t = 1, \tag{4.27}$$

and hence

$$\epsilon_t = x_t * b_t = q_t * a_{t\,min} * b_t = q_t. \tag{4.28}$$

The result of filtering x_t with b_t is to produce an error series ϵ_t that is exactly equal to the white-noise sequence q_t, as discussed earlier.
   b. If $a_t \neq a_{t\,min}$, the filter is not minimum-phase, and

$$a_t * b_t \neq 1. \tag{4.29}$$

In fact, the result of convolving a_t with b_t will be a series y_t, given by

$$y_t = \sum_{s=0}^{n+1} b_s \, a_{t-s}. \tag{4.30}$$

Thus

$$\epsilon_t = x_t * b_t = q_t * a_t * b_t = q_t * y_t, \tag{4.31}$$

or,

$$\epsilon_t = \sum_{s=0}^{\infty} y_s \, q_{t-s}. \tag{4.32}$$

The autocorrelation of y_t is $\phi_{yy}(\tau)$:

$$\phi_{yy}(\tau) = \phi_{bb}(\tau) * \phi_{aa}(\tau). \tag{4.33}$$

Now a_t and $a_{t\,min}$ have the same autocorrelation $\phi_{aa}(\tau)$, by definition. Therefore, comparing equations (4.33) and (4.27), we see that

$$\phi_{yy}(\tau) = 1, \quad \tau = 0$$
$$= 0, \quad \tau \neq 0. \tag{4.34}$$

The mean-square error I is given by

$$I = E\{\epsilon_t \cdot \epsilon_s\}$$
$$= E\{q_t * y_t \cdot q_s * y_s\}$$
$$= E\{q_t \cdot q_s\} * \phi_{yy}(t-s). \tag{4.35}$$

From equations (4.33) and (4.34) we see that

$$I = E\{q_t \, q_s\}$$
$$= \sigma_q^2, \quad t = s$$
$$= 0, \quad t \neq s. \tag{4.36}$$

Therefore, the mean-square error is σ_q^2, whatever the phase of q_t. It follows that the third limitation mentioned earlier does not alter the mean-square error. The exact values of the error series ϵ_t, however, will differ from the white-noise input q_t unless $a_t = a_{t\,min}$.

This is a significant result. Unless the filter a_t is minimum-phase, the error series, or *prediction error* ϵ_t, will not be equal to the white-noise sequence q_t. If b_t is the exact inverse of $a_{t\,min}$, however, whatever the phase of a_t, then the prediction-error series ϵ_t will have the same *statistics* as q_t. That is:

$$E\{\epsilon_t\} = E\{q_t\} = 0. \tag{4.37}$$

$$E\{\epsilon_t \, \epsilon_s\} = E\{q_t \, q_s\} = \sigma_q^2, \quad t = s$$
$$= 0, \quad t \neq s.$$

2. b_t is an approximate inverse of $a_{t\,min}$.
   a. If $a_t = a_{t\,min}$, we have

$$y_t = \sum_{s=0}^{n+1} b_s a_{t-s} \neq 1. \tag{4.38}$$

$$\epsilon_t = \sum_{s=0}^{\infty} y_s q_{t-s}. \tag{4.39}$$

Proceeding in the same way as in case 1a previously, we find

$$I = E\{\epsilon_t \, \epsilon_s\} = E\{q_t \, q_s\} * \phi_{yy}(t - s). \tag{4.40}$$

Because b_t is only an approximation to the inverse of $a_{t\,min}$, it follows that $\phi_{yy}(\tau)$ is not an impulse at $\tau = 0$. It is a smoother function with values for $\tau \neq 0$. Therefore,

$$I = E\{\epsilon_t \epsilon_s\}$$
$$= \sigma_q^2 \, \phi_{yy}(t - s), \tag{4.41}$$

and the error series ϵ_t is not white.
   b. If a_t is not minimum-phase, the mean-square error is the same:

$$I = \sigma_q^2 \, \phi_{yy}(t - s). \tag{4.42}$$

As in case 1 previously, the actual values of the error series ϵ_t will depend on the phase of the linear filter a_t. If $a_t = a_{t\,min}$, the error series will be a smoothed version of the input white-noise sequence q_t. If a_t is not minimum-phase, the error series will have the same statistical properties but will bear less resemblance to q_t locally.

4.6. Implications for Spiking Deconvolution of Seismic Data— An Example

The preceding section has direct implications for the spiking deconvolution process described in chapter 1. If the seismic trace x_t is the convolution of a wavelet s_t with the impulse response of the earth g_t, we see that we can find a spiking filter f_t to deconvolve the data and recover g_t, provided g_t is white, random, and stationary and provided s_t is minimum-phase. Even if g_t is white, random, and stationary, we cannot recover it from x_t unless s_t is minimum-phase. If s_t is not minimum-phase, the best that our Wiener spiking filter can do is to create a prediction-error series ϵ_t that has the same statistics as g_t but that differs from g_t locally.

This is a serious problem. It could mean that at a particular zone of interest the polarity of the recovered reflection coefficient series is wrong and that, moreover, a big reflector above a little one might appear as a small reflector above a big one, or vice versa. Locally, then—just where the phase is very important—the picture can be distorted. The broad picture, however, will be correct. This is not very comforting to the seismologist who is looking for the very detail that the deconvolution process was meant to reveal. If the details cannot be trusted, what good is the process? This is a good question.

To get some idea of the errors involved, we consider a synthetic example that includes the problem of spherical divergence and a non-minimum-phase wavelet. The problems of a nonwhite, nonrandom geological sequence are not considered. Neither is the influence of noise.

In our example we begin with a white, random, stationary sequence $\hat{r}\ (t)$, as shown in figure 4.9(1). This sequence was constructed in the frequency domain by making the amplitude spectrum constant and the phase spectrum random for positive frequencies such that the probability of the phase having a value between θ and $\theta + \Delta\theta$ was $\Delta\theta/2\pi$. For negative frequencies the phase was made the negative of the phase for the corresponding positive frequencies, thus ensuring that the time-domain sequence would be real, as shown in figure 4.9(1).

Because of spherical divergence, the impulse response of the layered elastic earth cannot be white, random, and stationary, as discussed in chapter 2. We have simulated this effect by applying an inverse spherical-divergence correction to our white, random, stationary sequence. That is, we divide $\hat{r}\ (t)$ by the function bt^n to create a sequence

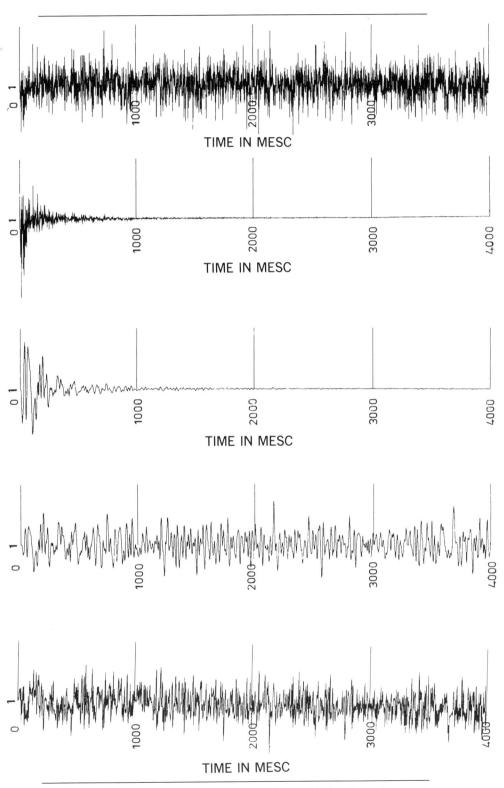

Figure 4.9. 1. White random stationary noise sequence. 2. Sequence (1) after spherical divergence. 3. Sequence (2) after convolution with wavelet of figure 4.11(a). 4. Sequence (3) after spherical divergence correction. 5. Sequence (4) after one-step-ahead prediction-error filtering.

$$r(t) = \frac{\hat{r}(t)}{bt^n},$$
(4.43)

which is shown as figure 4.9(2). This is now the synthetic impulse response of the earth.

Next we wish to convolve this impulse response with a wavelet. We have taken as our initial wavelet the measured air-gun array signal shown in figure 4.10. This was measured at a depth of 100 m below the array over the bandwidth 0–125 Hz with a steep cutoff of 72 dB per octave above 125 Hz. (This Texas Instruments DFS V filter is known to be minimum phase.) In normal processing, frequencies much above 40 Hz are seldom retained on the final display, so we have filtered the wavelet of figure 4.10 with a minimum-phase filter to attenuate frequencies above this frequency. The result is shown in figure 4.11(a). In figure 4.11(b) we show the minimum-phase wavelet, which has the same power spectrum as the wavelet in 4.11(a). Clearly, the measured minimum-phase filtered wavelet is not minimum-phase. It follows that the original measured wavelet of figure 4.10 is also not minimum-phase.

We next convolve the wavelet of figure 4.11(a) with the synthetic earth impulse response [figure 4.9(2)] to produce the sequence shown in figure 4.9(3). This is now a noise-free point-source synthetic seismic trace. We will now try to recover both the wavelet and the original white, random, stationary sequence from trace 4.9(3).

The first step in this recovery is to apply a spherical-divergence correction to make the data stationary. That is, we wish to change our point-source data such that they appear to satisfy a plane-wave model. Normally there are multiples and noise to confuse the problem. Furthermore, the spherical-divergence correction is normally not known: it must be guessed. In this case multiples and noise are absent, and we know the spherical-divergence correction exactly: it is the function, bt^n. We can make a "perfect" spherical-divergence correction by multiplying the synthetic seismic trace of figure 4.9(3) by this function. The result is shown in figure 4.9(4) and would seem to be reasonably stationary. We know, however, that the price we have paid for achieving this stationarity is distortion of the wavelet, as described in section 2.7. We shall see shortly how much distortion has been introduced.

Now that the data are stationary, we may apply our statistical deconvolution technique, assuming, first, that the impulse response of the earth is white, random, and stationary, and, second, that the wavelet is minimum-phase. Our first assumption is as good as we can make it, given that we have to force point-source data to have the property of stationarity. The second assumption is not so good but is at the same level of accuracy as most standard marine seismic processing carried out at present (1983). With our first assumption holding fairly well, we can confidently compute an autocor-

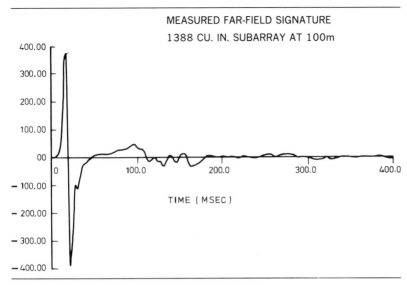

Figure 4.10. Far-field signature of an air-gun subarray.

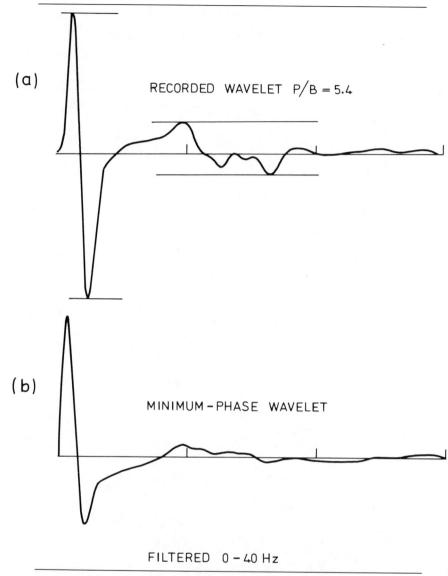

(a)

RECORDED WAVELET P/B = 5.4

(b)

MINIMUM-PHASE WAVELET

FILTERED 0 – 40 Hz

Figure 4.11. (a) Minimum-phase low-pass filtered recorded wavelet of figure 4.10. (b) Minimum-phase wavelet with same spectrum as wavelet (a), but not displayed with same amplitude scale.

relation function for the data of figure 4.9(4) and compute a one-step-ahead prediction-error filter as described in section 4.3. This is the same as spiking deconvolution, as described in section 4.4, provided the earth impulse response is white, random, and stationary. The result of this prediction-error filtering of the trace 4.9(4) is the trace 4.9(5). If we have done everything correctly, 4.9(5) should be similar to 4.9(1).

In figure 4.12 we show these two traces—the original white-noise sequence 4.9(1) and the deconvolved synthetic seismogram 4.9(5)—displayed close together so that their similarities and differences can be readily appreciated. What we see is two random sequences that look as if they probably have the same low frequency spectral components. Sequence 4.9(1) has greater higher frequency content than 4.9(5). The high frequencies were lost in the convolution with the low frequency air-gun wavelet. Deconvolution cannot restore the frequencies which are missing. Sometimes the two sequences are in phase, and sometimes they are not. In section 4.5 we predicted that the deconvolved trace or error series would have the same statistical properties as the original white-noise sequence, but would differ from it locally because the wavelet is not minimum-phase. This is seen to be broadly true, but because of the low pass filtering effect of the convolution with the wavelet, we cannot even achieve the same statistical properties as the input sequence.

The minimum-phase wavelet that has the same autocorrelation as the filtered measured wavelet is shown in figure 4.11(b). Because of the spherical-divergence correction, some nonlinear distortion is introduced into the wavelet. The prediction-error filter is thus the inverse of some other wavelet. By finding the inverse of the prediction-error filter by least-squares, we can find the expected minimum-phase wavelet as described at the end of section 3.7. The recovered wavelet and the minimum-phase equivalent of the input wavelet are shown in figure 4.13. Quite clearly the expected wavelet and the original input wavelet are different.

We shall refer to this example in later chapters. The main conclusion is that errors in the estimation of the wavelet have serious consequences for the results of the deconvolution, even if all the statistical assumptions are valid. The problem of spherical divergence makes matters worse.

4.7. Is the Wiener Prediction Filter Unique?

In section 4.2 we remarked that equations (4.3) yield a single prediction filter even though there are many sequences that possess the same autocorrelation function. There is a little mystery to be unravelled here. In section 4.3 we computed a prediction-error filter directly from the prediction filter. Therefore, the least-squares technique gives a unique prediction-error filter for a given autocorrelation function, prediction distance, and filter length.

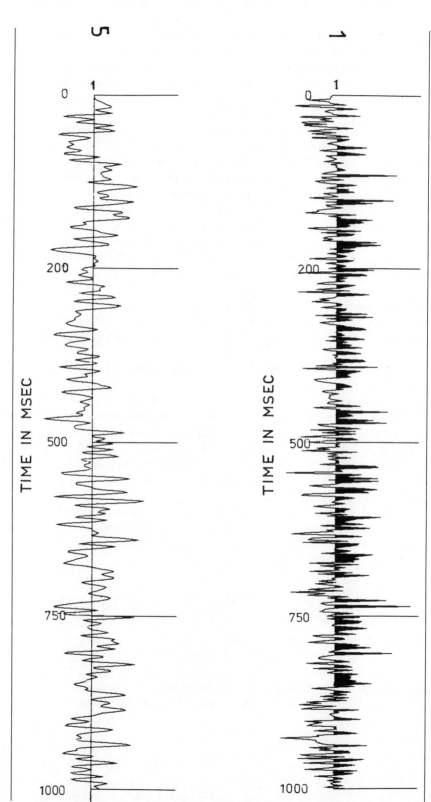

Figure 4.12. Enlargements of sequences 4.9(1) and 4.9(5).

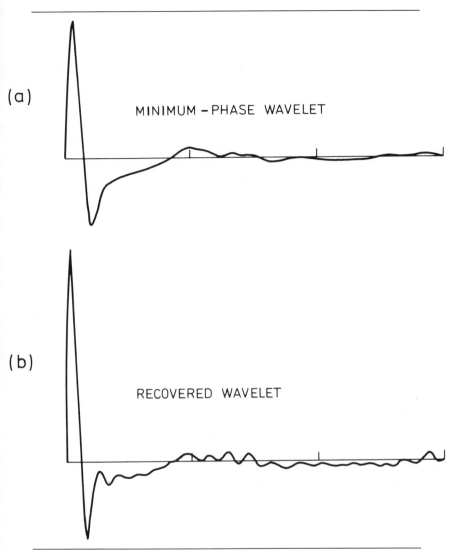

(a)

MINIMUM – PHASE WAVELET

(b)

RECOVERED WAVELET

Figure 4.13. Distortion introduced by perfect spherical-divergence correction. (a) Minimum-phase equivalent of input signal wavelet. (b) Recovered minimum-phase wavelet after application of spherical divergence correction.

The least-squares one-step-ahead prediction-error filter was shown in section 4.4 to be exactly the same as the least-squares inverse of the minimum-phase smoothing filter. If the smoothing filter is not minimum phase the best least-squares inverse will have a different phase spectrum from this prediction-error filter, although the result of convolving the non-minimum-phase smoothing filter with the best least-squares inverse may now delay the spike (see section 5.1). The error energy will be the same provided the autocorrelation of the applied filter remains the same as that of the prediction-error filter, as shown in section 4.5.

We can now consider the suite of filters which has the autocorrelation of the one-step-ahead prediction-error filter. If the smoothing filter of our white random stationary noise sequence is non-minimum-phase, we should find its best least squares inverse of given length in this suite. We can think of this least-squares inverse as a kind of prediction-error filter for the non-minimum-phase smoothing filter, to which there should correspond some prediction filter.

Let b_t be a one-step-ahead prediction-error filter with the same autocorrelation $\phi_{bb}(\tau)$ as $b_{t\min}$, the prediction-error filter determined from the normal equations (4.25). Following the analysis of section 4.5, we may write down the mean-square error I from equation (4.42) as:

$$I = \sigma_q{}^2 \, \phi_{yy}(t - s),\qquad(4.42)$$

where $\phi_{yy}(\tau)$ is given by equation (4.33):

$$\phi_{yy}(\tau) = \phi_{bb}(\tau) * \phi_{aa}(\tau).\qquad(4.33)$$

It follows that the mean-square error I is the same, independent of the phase of the prediction-error filter b_t.

We recall from section 4.4 that, from equations (4.20):

$$b_{0\ \min} = 1.$$

$$b_{i\min} = -p_{i-1}, \quad i = 1, 2, \ldots, n + 1.\qquad(4.44)$$

Now, because $b_{t\min}$ is minimum-phase, we know that:

$$b_{0\min} > b_0 \, ,\qquad(4.45)$$

where b_0 is the first coefficient of b_t, the non-minimum-phase prediction-error filter. This follows from Robinson's energy-delay theorem. Thus, combining equations (4.44) and the inequality (4.45), it follows that:

$$b_0 < 1.\qquad(4.46)$$

From equation (4.11) we see that the z-transform $B(z)$ of the prediction-error filter b_t is given by:

$$B(z) = 1 - z^{\alpha} P(z)$$

$$= b_0 + b_1 z + b_2 z^2 + \ldots + b_{n+1} z^{n+1}, \tag{4.47}$$

where α is the prediction distance, and $P(z)$ is the z-transform of the prediction filter corresponding to the prediction-error filter. [Equation (4.47) is simply a consequence of the definition of the error sequence. It has nothing to do with least squares.] Rearranging equation (4.46), we have:

$$P(z) = z^{-\alpha} [1 - B(z)]$$

$$= z^{-\alpha} (1 - b_0 - b_1 z - \ldots - b_{n+1} z^{n+1}). \tag{4.48}$$

In the case where α is 1, we know that the coefficients of the least-squares prediction-error filter obey equations (4.44) and that, because this filter is the same as an inverse filter, it must be minimum-phase. There is only one minimum-phase wavelet in any suite of wavelets that share the same autocorrelation function. All the other wavelets in the suite are not minimum-phase, and therefore the first coefficient in any of the other wavelets in the suite obeys the inequality (4.46). It follows that the prediction filter corresponding to any non-minimum-phase one-step-ahead prediction-error filter must have a z-transform with a term $(1 - b_0)z^{-1}$. This means that there will be a coefficient in the corresponding one-step-ahead prediction filter. This is an *anticipation* component and would make p_t noncausal. Wiener filters can only be causal, by definition.

Thus we see that there is a suite of prediction-error filters corresponding to our least-squares minimum-phase one-step-ahead prediction-error filter. To each one of the prediction-error filters in this suite there is a corresponding prediction filter. All except one of these filters is noncausal, however. The only causal one is the least-squares Wiener prediction filter.

This result, for one-step-ahead prediction, can be extended to the more general case of least-squares prediction for any prediction distance α. (See Berkhout and Zaanen 1976.) As we saw in section 4.3, the whole process of computing a prediction-error filter can be regarded as that of finding an inverse of the predictable part of the data. The purpose of prediction-error filtering is to remove that predictable part. Whether the predictable part is removable or not depends on whether or not it is minimum-phase.

4.8. Implications for the Removal of Multiples and Bubble Pulses

Typically, the seismic wavelet is not minimum-phase: for example, the airgun signal of figure 4.11(a) is not minimum-phase. Nevertheless, prediction-error filtering can sometimes be successful if long-period multiples are present. Consider, for example, the situation shown in figure 4.14, where our geological sequence contains a thick layer near the surface bounded at both top and bottom by strong reflectors. Seismic energy will tend to get trapped in this layer, but energy can penetrate through the layer in both directions.

If we now consider a reflection from a deeper interface below this thick layer, we shall see, first, a primary reflection P corresponding to the arrival of the wave that has traveled straight down to this interface and back up to the receiver. This primary will then be followed by a complicated sequence of multiples caused by extra bounces of the seismic wave in the thick layer, as shown in figure 4.14. These extra bounces are experienced by both the downgoing incident wave and the returning reflected wave. Thus the primary reflection is followed by a succession of echoes corresponding to these additional multiple reflections. If the two-way travel time to the reflector is t_1, and the two-way travel time in the thick layer is T_1, we should expect to see arrivals at times t_1, $t_1 + T_1$, $t_1 + 2T_1$, and so on.

If we now consider another deep reflector with two-way travel time t_2, we should expect its primary arrival at time t_2 to be followed by a similar sequence of multiples such that the arrival times of the primary and corresponding multiples from this reflector would be t_2, $t_2 + T_1$, $t_2 + 2T_1$, and so on. Any primary reflected wave from below this layer must pass through the layer twice. Successive corresponding multiple reflections must pass through the layer four times, six times, eight times, and so on. The actual sequence of multiples depends only on the characteristics of the layer. In fact, it depends on the *two-way transmission response* of the layer, which we can denote by $m(t)$. The part of the seismogram that represents reflections from below the layer can be thought of as the convolution of the primary reflection with this two-way transmission response $m(t)$. If we denote the seismic wavelet by $s(t)$, the sequence of primary reflections by $g(t)$, and the received seismogram by $x(t)$, then we have

$$x(t) = s(t) * g(t) * m(t). \tag{4.49}$$

This equation says that *every* primary event in the sequence $g(t)$ is followed by a sequence of multiples $m(t)$, and that every primary and every multiple reflection is represented by the same wavelet $s(t)$. For this convolutional representation to make sense, *every* primary reflection in the sequence $g(t)$ must come from a reflector *below* the layer that generates the multiples. Therefore, the layer must be very close to the surface; specifically it must be above all other reflecting interfaces. If the layer is not close to the surface, the primary reflections from above the layer will not be followed by the multiple sequence that follows any primary reflection from below the layer, and the convolutional equation (4.49) will not apply.

Let us assume that the layer is close to the surface. If we have a point source separated from the receiver, as shown in figure 4.14, it is a simple consequence of geometry that the two-way transmission response of the

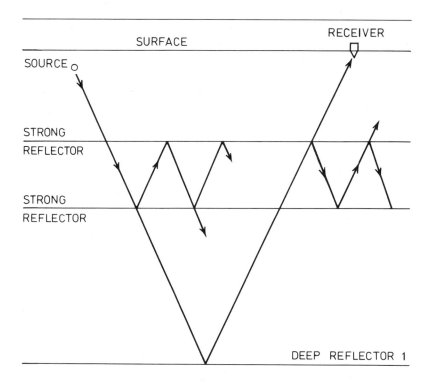

Figure 4.14. The effect of a multiple-generating layer.

layer cannot be the same for any two deeper reflectors (see figure 4.15). Equation (4.49) says that this two-way transmission response *is* the same for every primary reflection. In fact, equation (4.49) can hold only for the one-dimensional plane-wave model discussed in section 2.5. Thus, in this one-dimensional time-domain representation, the multiple sequence $m(t)$ is not strictly convolutional.

Let us ignore this difficulty for the moment, and see how predictive deconvolution could be used to predict out long-period multiples, on the basis that equation (4.49) is reasonably accurate.

The autocorrelation of $x(t)$ is

$$\phi_{xx}(\tau) = \phi_{qq}(\tau) \ast \phi_{ss}(\tau) \ast \phi_{mm}(\tau). \tag{4.50}$$

In figure 4.16 we show what this autocorrelation would look like if $g(t)$ were white, random, and stationary and if the autocorrelation $\phi_{ss}(\tau)$ existed only for lags τ less than $T_1/2$, where T_1 is the two-way travel time in the thick layer. The individual autocorrelations $\phi_{qq}(\tau)$, $\phi_{ss}(\tau)$, and $\phi_{mm}(\tau)$ are also shown. In a situation exactly like this, the prediction problem and prediction-error problem for the multiple sequence would be straightforward.

We could arrange for the prediction distance α to be greater than the maximum lag of the autocorrelation of the seismic wavelet, and the length of the filter to be equal to $\alpha + n$, as shown in figure 4.16. The prediction-error filter would then operate as shown in figure 4.6. It would use samples before and after time t to predict and subtract a multiple at time $t + T_1$. If the multiple sequence is minimum-phase, the least-squares prediction-error filter can effectively attenuate the multiples and thus reveal the primary reflections (see section 4.7). Treitel and Robinson (1966) showed that, for the one-dimensional plane-layered earth, the multiple sequence generated by an impulsive plane wave *is* minimum-phase. Thus there is good reason to expect least-squares predictive deconvolution to have some probability of success.

Note that in this case, where we do not have any overlap between the autocorrelation of the seismic wavelet and the multiple sequence, we do not also have to assume that the wavelet is minimum-phase. Provided we choose a predictive gap larger than the length of the wavelet, we can deal with the multiples as a separate problem.

Very often, however, the autocorrelation of our data is not as easily interpreted as in figure 4.16. Usually there is more than one sequence of multiples. Thus we would have $m_1(t)$, $m_2(t)$, $m_3(t)$, and so on, each of which we should like to regard as convolutional. With marine data there are very often *bubble pulses* in the signal $s(t)$, and these give rise to secondary and tertiary peaks in the autocorrelation function that may well overlap with the peaks caused by the multiples. There is then a problem in interpreting the structure of the autocorrelation function: Which peaks are which?

In other words, if the wavelet $s(t)$ is not known, and its autocorrelation

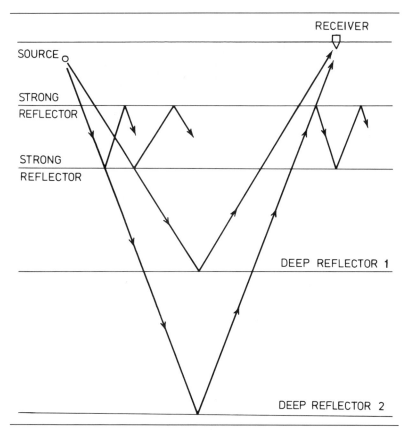

Figure 4.15. The two-way travel time in the multiple-generating layer depends on the angle of incidence. Therefore the multiples are not strictly periodic.

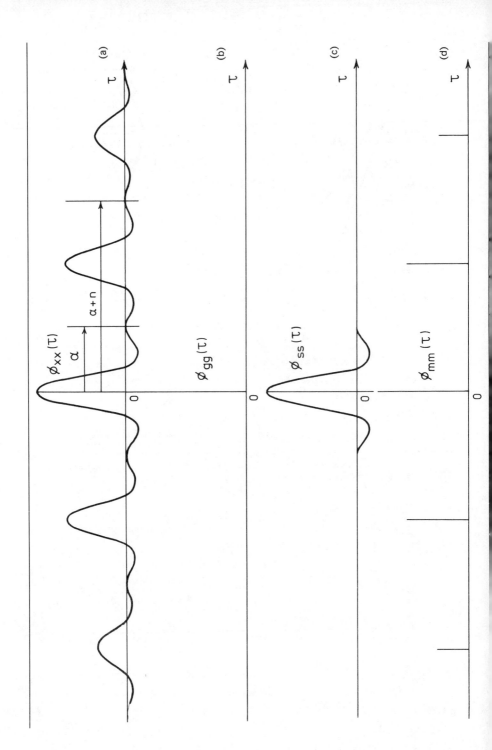

$\phi_{ss}(\tau)$ is not known, how can the minimum-phase multiple sequence be separated from the non-minimum-phase bubble-pulse sequence? Of course this separation cannot be made. Moreover, because the bubble-pulse sequence is *not* minimum-phase, the prediction-error filter cannot work properly.

If, on the other hand, the wavelet *were* minimum-phase, there would be no need to separate the wavelet from the multiples: the prediction-error filter would work properly anyway. The problem is to determine whether or not the wavelet is minimum-phase. This cannot be done without first knowing what the wavelet is.

4.9. The Problem of Estimating the Autocorrelation Function: Seismology versus Econometrics

In the Appendix we define the autocorrelation of a seismogram x_t as:

$$\phi_{xx}(\tau) = \sum_{t=0}^{\infty} x_t x_t - \tau. \tag{4.51}$$

In practice, we do not record forever; thus we have only a finite number of data points, say $N + 1$. The upper limit of the summation is therefore restricted by the length of data, and the autocorrelation must be redefined as:

$$\phi_{xx}(\tau) = \sum_{t=\tau}^{N} x_t x_{t-\tau}. \tag{4.52}$$

This is illustrated in figure 4.17. Notice that the zero-lag coefficient $\phi_{xx}(0)$ is calculated from $N + 1$ terms, $\phi_{xx}(1)$ is calculated from N terms, and so on. If N is not very large compared with the maximum lag of interest, this will have an undesirable biasing effect at large lags. The biasing is removed if the summation is normalized by dividing by the number of terms in the summation:

$$\phi_{xx}(\tau) = \frac{1}{N - \tau + 1} \sum_{t=\tau}^{N} x_t x_{t-\tau} \tag{4.53}$$

Although equation (4.53) solves the problem of the end effect, it creates another problem: it does not necessarily give a valid autocorrelation function. The Fourier transform of an autocorrelation function is the power spectrum, which is real and positive for all frequencies. Using equation (4.53) on certain data samples results in functions whose Fourier transforms are not positive at all frequencies; that is, these functions are not autocorrelations, and their Fourier transforms are not power spectra.

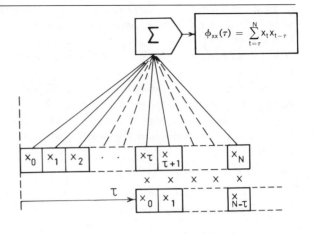

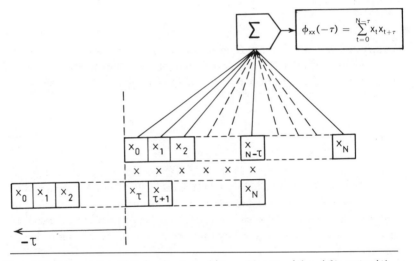

Figure 4.17. Autocorrelation of finite-length sequence of data x_t*: (a) positive shift; and (b) negative shift.*

This problem is solved with the *maximum entropy method* of spectral analysis. In geophysics this method is always associated with the name of Burg (1967, 1972, 1975). An excellent explanation of Burg's method and a FORTRAN computer program is given by Claerbout (1976, pp. 133–137). The essence of Burg's approach is to assume that the data in the time window 0–N are representative of a much longer sequence, existing before $t = 0$ and after $t = N$. That is, the whole sequence is regarded as stationary. A special prediction-error filter is calculated directly from the data and is forced to be minimum-phase. The computation of the filter is recursive, beginning with one filter coefficient, then two, then three, and so on. The design criterion for the filter is that the sum of the squares of the errors of both the one-step-forward and the one-step-backward prediction are minimized for each step in the recursion. As the filter length increases, the prediction-errors become unpredictable random numbers, or white noise.

Since the convolution of the Burg prediction-error filter with the data is a white noise, the power spectrum of the filter is the inverse of the power spectrum of the data. The filter is of finite length, and its power spectrum can be calculated exactly from its Fourier transform. The inverse of this power spectrum is the power spectrum of the data, and the Fourier transform of the power spectrum of the data is the required autocorrelation function. Therefore, provided the data are stationary, Burg's maximum entropy method can provide an estimate of the autocorrelation of the data, which is not affected by the end effects of equation (4.53). It achieves what, at first sight, equation (4.54) apparently ought to achieve, but does not.

At large lags there is still a problem of statistics. In order to estimate the influence of some periodic feature in the data, the autocorrelation must be computed for lags at least as big as the period of this feature. If there are insufficient cycles of this periodic feature in the data, it will not be possible to estimate the strength of this feature with much confidence. Typically about 10 cycles are normally required. That is, the lag τ should not be made much more than $N/10$ if some confidence is to be placed in the spectral estimate. Of course, an estimate *can* be made even if τ exceeds $N/10$; the problem is to decide how much confidence to place in it. The statistics can be improved by averaging autocorrelations across a shot record, but because the geometry of each source-receiver pair is different, it is not possible to average estimates of the same thing.

It should be noted that the application of some sort of spherical-divergence correction to the raw field data may not be sufficient to make it stationary. Figure 4.18 shows a stacked section of a seismic line with horizontal bedding down to an unconformity at about 0.3 seconds. Below the unconformity the beds are dipping. Associated with this unconformity, which is a *geological feature,* is a change in the spectral bandwidth of the data. Below the unconformity the high frequencies are rapidly attenuated, and only low-frequency energy is apparent on the section. The filtering of this section has been uniform from top to bottom. The change in spectral bandwidth at the unconformity must therefore be caused by the geology rather than by data processing.

The data are clearly not stationary, having a broader bandwidth above the unconformity than below it. We can try to apply the concept of stationarity by dividing the data into two parts at the unconformity: we may then look at the statistics of the data in these two time windows separately.

In order to investigate the statistics in these two separate time windows, we must regard each window as representative of some longer stationary sequence. It is quite possible that the windows are now not long enough to provide adequate statistics at long lags. If the windows are not about ten times longer than the longest lag of interest, there is a significant probability than an accidentally high correlation between two primary reflections will appear in the autocorrelation and be interpreted as a multiple. What we do in this situation depends on how serious we consider the long-period multiple problem to be. If the problem is serious enough, and if predictive deconvolution is the only way available to us to suppress the multiples, then we must compromise. We shall have to sacrifice our principles about the spectral stationarity and simply ensure that the window over which the autocorrelation is to be estimated is long enough to permit a good estimate of the long-period prediction-error filter to be made.

Seismic data are different from many other kinds of data. Before the shot is fired, there is only noise. Zero time on seismic data always refers to the instant the shot is fired. If we want to increase our window length, we can do so only if we have recorded for long enough. For example, if we have a multiple of 200 ms period, we should need to record for about 2 seconds. If our target is at 400 ms, however, a 2 second record length seems excessive. This situation is illustrated in figure 4.19. If predictive deconvolution is the way to attack the troublesome long-period multiple problem and reveal the primary reflection at 400 ms, it can be done only if a 2 second record length was used at the acquisition stage. Data processing problems such as this need to be discovered before too many data are recorded with the wrong parameters.

Even a long record length does not really solve the problem, however. There is a fundamental problem for seismic data that cannot be encountered in, say, econometric data. The prediction-error filter as described in section 4.3 predicts and subtracts the *predictable part* to uncover the *unpredictable part*. In the example of figure 4.18, we would like our prediction-error filter, calculated from the autocorrelation function, to predict out the multiple at 400 ms and uncover the primary reflection at the same time. The primary reflection is regarded as unpredictable. It is not *inherently* unpredictable, however, because it is certain to be included in the window we will use to

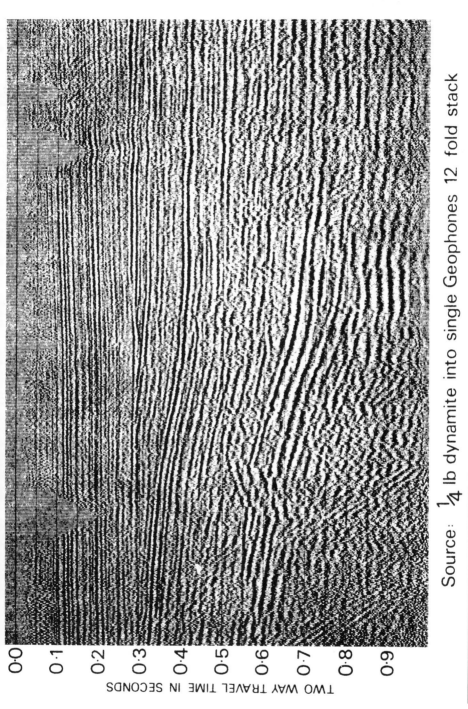

TWO WAY TRAVEL TIME IN SECONDS

Source: ¼ lb dynamite into single Geophones 12 fold stack

Figure 4.18. Stacked section showing nonstationarity of stacked seismic data. Above unconformity at 0.3.seconds the data have higher frequency content than below 0.3 secs. Reprinted by permission of Geophysical Prospecting from Ziolkowsk: and Lerwill 1979, fig. 25, p. 389.

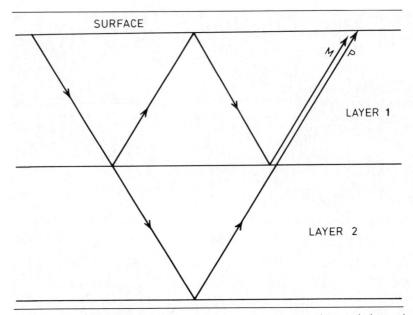

SURFACE

LAYER 1

LAYER 2

Figure 4.19. The two-way travel time in layer 1 is 200ms. The one-way travel time to the bottom of layer 2 is 400ms. The primary reflection from the bottom of layer 2 and the first multiple reflection within layer 1 arrive at the same time.

estimate the autocorrelation function. It will therefore contribute to our estimate of the predictable part. The only way we could avoid including the primary at 400 ms is to choose a window that stops before 400 ms; but we cannot do that because the window will then be too short: there are no data before zero time.

In econometrics this problem does not arise. Predictions about the future behavior of the economy are based on data from the past and present. Events that have yet to happen cannot bias the estimates we make today. In seismic data processing the terms *past, present,* and *future* do not have the same meaning as they do for econometricians, because all the seismic events have already been recorded. Econometricians can base their predictions on data that go back as far as they can be trusted to be stationary. Seismologists have highly nonstationary data and can go back only to the instant the shot is fired. In order to get enough statistics for reliable prediction, it is often esential to use a window that includes the very "unpredictable" event that is to be revealed.

In practice, the only way to minimize the effect of this unpredictable part on the estimate of what is predictable is to use very long windows. If short windows are used, there is always the risk that an accidentally large correlation between two primary reflections will occur at the same lag as the multiple and thus bias the estimate. The prediction-error filter calculated from such a biased autocorrelation function will then act to *predict and subtract out* the second primary reflection. In other words, the filter will remove the very geological information it is intended to reveal. The issue comes down to how well or not $\phi_{gg}(\tau)$ approximates an impulse function. If the primary sequence is random over a given window, that window need not avoid the "inherently unpredictable" primaries. The problem is to establish that $\phi_{gg}(\tau)$ is indeed random over any given window.

4.10. Aside: The Effect of Normal Moveout on Poststack Deconvolution

From the considerations that have been described so far, it is clear that if deconvolution of the source signature is to be attempted, it should be done before any spherical-divergence correction is applied. It is also clear that this signature deconvolution cannot be attempted so early in processing unless the source wavelet is known.

Suppose the wavelet is not known. Is there some optimum processing sequence that will allow the wavelet to be estimated accurately? As we shall see in section 6.4, in my view there is no scientific way to answer this question. Nevertheless, there is one fairly popular processing sequence that attempts to bring the data into shape for wavelet extraction at the poststack stage, which we shall discuss here. The sequence is as follows:

1. Demultiplex.
2. Spherical-divergence correction.
3. Predictive deconvolution (gap deconvolution).
4. Normal moveout correction.
5. Stack.
6. Wavelet extraction.
7. Signature deconvolution.

We have already mentioned in several places that the spherical-divergence correction violates the convolutional model and distorts the wavelet unevenly down the seismic trace. Earlier in this chapter, however, we have shown that this distortion is a price that is often willingly paid in order to force the data to be stationary.

The predictive-deconvolution step is usually a fairly short prediction-error filter in which the prediction distance or gap α is only a few samples. This filter is intended to remove some of the short-period multiples introduced by, for example, the near-surface layering. The problem with this filter is that there is very likely to be some overlap between the autocorrelation of the wavelet and the autocorrelation of the short-period multiples, and then the filter cannot work properly if the wavelet is not minimum-phase, since the predictable part must be minimum-phase (see section 4.8)

It is ironic that the reason for applying this filter at this stage is to clean up the data in order to make a better estimate of the wavelet later on in processing. It is quite likely that this predictive deconvolution filter, which is calculated on a trace-by-trace basis and is therefore data-dependent, distorts the wavelet in a different way on each trace. This distortion is added to the distortion that has already been introduced by the spherical-divergence correction.

In normal seismic data acquisition, the common depth point—or, more correctly, common midpoint—system is used (Mayne 1962). There are usually many source-receiver pairs that share the same common midpoint, and usually the incremental separation or "offset" between successive source-receiver pairs is a constant distance, as shown in figure 4.19(a). The shot-receiver pairs that share the same common midpoint are gathered together to form a *common midpoint gather,* as shown in figure 4.19(b), in which the traces are shown displayed side by side in order of increasing offset.

When the reflecting horizons are parallel to the surface, as shown here, the travel time for a primary reflection increases in a nonlinear way with increasing offset because the horizontal component of the travel time increases while the vertical component stays the same. This effect is called *normal moveout.* As the depth of the reflectors increases, the effect of normal moveout decreases because the horizontal component of the travel time becomes a smaller and smaller fraction of the total travel time.

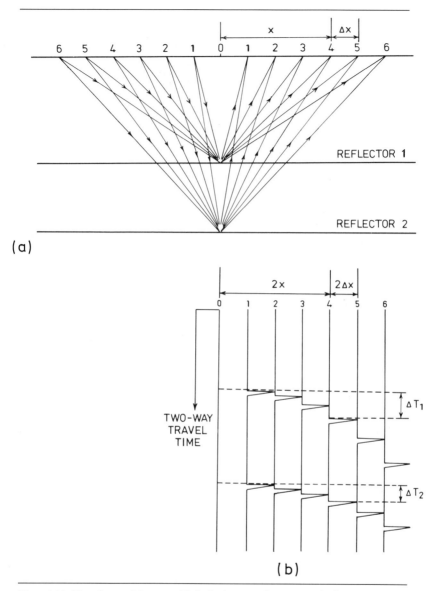

Figure 4.20. Normal moveout decreases with depth of target and increases with offset.

In a common midpoint gather this normal moveout effect can be corrected by applying time shifts to the individual reflections to make them appear as if they had occurred for zero offset on all traces of the gather. As can be seen from figure 4.20, these time shifts are not constant along each trace: and on the far-offset traces the shallow reflections are shifted far more than the deeper reflections. This causes the data to be stretched.

In fact, of course, we do not have individual reflections. We have primary reflections and multiples, and every one of these events is represented by some (distorted) wavelet. Dunkin and Levin (1973) pointed out that the normal moveout correction stretches the wavelet unevenly. A wavelet arriving early is stretched more than one arriving late at the same offset, and wavelets arriving from the same reflector are stretched increasingly with increasing offset. In other words, the normal moveout correction introduces two kinds of uneven nonlinear distortion to the wavelet: both *down* and *across* the gather.

After normal moveout correction, the traces are summed or stacked. The stacked trace then contains a mixture of all these wavelets, which have now been distorted in at least four different ways: by spherical-divergence correction, by predictive deconvolution, and by normal moveout stretch both down and across the gather.

At this point in this processing sequence, the stacked trace is considered to have been massaged into shape for "the wavelet" to be extracted. The so-called wavelet actually consists of a large number of wavelets, each of which began its life as a signal generated by the source and has subsequently been distorted in various ways. Whatever is done to "the wavelet" that is extracted at this point in processing relies more on luck than on any clear relation to the convolutional model of the seismogram.

5. SIGNATURE DECONVOLUTION

5.1. Spiking Deconvolution of a Known Wavelet

Sometimes the wavelet is known from measurements. These measurements are far easier to make for marine sources than for land sources. The exact shape of the wavelet generated by a land source is usually extremely difficult to determine. On the other hand, with the exception of Vibroseis*, the source mechanism is usually very impulsive and the source wavelet is fairly short and sharp, as required. Marine sources are usually much less impulsive and are usually either explosive or implosive. An *ex*plosive source, such as dynamite or an air gun, suddenly creates a high-pressure expanding gas bubble in the water. The motion of this bubble generates an oscillating sound wave that is usually not minimum-phase. An *im*plosive source, such as a water gun or a steam gun, creates a low-pressure cavity in the water that contains only water vapor. The collapse of the cavity generates a large impulse followed by subsequent smaller oscillations. The wavelet generated by an implosive source is always non-minimum-phase.

In general, with marine sources the wavelet is much less impulsive than the wavelet of a typical land source, and it is usually not minimum-phase. It may be estimated from measurements or even measured directly, however. How do we design a filter to turn such a wavelet into a spike?

From chapter 3 we know that such a filter must be the inverse of the wavelet. Let the wavelet be discretely sampled, with coefficients

$$s_t = s_0, s_1, \ldots, s_p. \tag{5.1}$$

The z-transform of this wavelet is

$$S(z) = \sum_{t=0}^{p} s_t z^t, \tag{5.2}$$

which can be factorized into q roots that lie outside the unit circle and $p - q$ roots inside the unit circle. Thus:

*Trademark of Continental Oil Company

$$\frac{S(z)}{s_p} = [\prod_{k=1}^{q} (z - z_k)] \cdot [\prod_{l=1}^{p=q} (z - z_l)],$$

where

$$|z_k| > 1, \quad k = 1, 2, \ldots, q,$$

and

$$|z_l| < 1, \quad l = 1, 2, \ldots, p\text{-}q. \tag{5.3}$$

The product of the q roots outside the unit circle is the z-transform of a *minimum-phase* wavelet, $s_{t\,min}$. The product of the $p-q$ roots inside the unit circle is the z-transform of a *maximum-phase* wavelet $s_{t\,max}$. Thus the wavelet s_t can be regarded as the convolution of a minimum-phase wavelet with a maximum-phase wavelet. Such a wavelet is known as a *mixed-phase* wavelet.

The inverse of s_t may be computed via the z-transform $S(z)$. Let the inverse of s_t be a_t. Then the z-transform of a_t is:

$$A(z) = \frac{1}{S(z)}$$

$$= \frac{s_p}{[\prod_{k=1}^{q} (z - z_k)][\prod_{l=1}^{p-q} (z - z_l)]}$$

$$= \sum_{t=-\infty}^{\infty} a_t z^t. \tag{5.4}$$

The coefficients $a_0, a_1, a_2, \ldots$ form a converging series of finite energy which is the causal inverse of the minimum-phase part of s_t. The coefficients $a_{-1}, a_{-2}, a_{-3}, \ldots$ form a converging series of finite energy which is the purely *anti*causal inverse of the maximum-phase part of s_t. As we know from chapter 3, both series are infinitely long.

An approximation of this noncausal inverse may be made with a finite-length filter that has an anticausal as well as a causal part. Let this approximate filter be b_t, with the $n + 1$ coefficients:

$$b_{-m}, b_{1-m}, \ldots, b_{-1}, b_0, b_1, \ldots, b_{n-m},$$

$$\uparrow \qquad\qquad \uparrow \qquad\qquad \uparrow$$

$$t = -m \qquad\qquad t = 0 \qquad\qquad t = n - m$$

such that there are m coefficients before $t = 0$ and $n - m$ coefficients after $t = 0$. The result of convolving b_t with s_t will be an approximation to a unit impulse at time $t = 0$. That is:

$$\sum_{k=-m}^{n-m} b_k \, s_{t-k} = d_t \backsimeq \delta_t, \tag{5.5}$$

where δ_t is the Kroneker delta:

$$\begin{aligned} \delta_t &= 1, \quad t = 0 \\ &= 0, \quad t \neq 0. \end{aligned} \tag{5.6}$$

The wavelet s_t is causal and does not exist before $t = 0$. The filter b_t is noncausal, and therefore d_t is noncausal and does exist before $t = 0$:

$$d_{-m} \, , \, d_{1-m} \, , \, \ldots , \, d_{-1} \, , \, d_0 \, , \, d_1 \, , \, \ldots .$$

$$\uparrow \qquad\qquad\qquad\qquad \uparrow$$

$$t = -m \qquad\qquad\qquad t = 0$$

The situation is illustrated in figure 5.1.

Mathematically, there is no difficulty in describing the operation of this noncausal filter. The problem is physical: the filter *anticipates* the input. Obviously we cannot design such a thing: it gives an output before the input has been applied. If such a thing existed, however, we could imagine it coupled to a tape recorder that starts recording the output d_t in anticipation of the input s_t. If we played back the tape, we would find the series d_t recorded in just the same way as if the system had been causal.

In geophysics we are nearly always dealing with recorded data. What matters is the numbers on the tape and the time origin. The numbers on the tape would be exactly the same if we chose a different time origin. The time origin would affect only our time-indexing system. Thus, if we chose a time origin such that our sample indices were denoted by τ, where

$$\tau = t + m, \tag{5.7}$$

the convolution of equation (5.5) would then be:

$$\sum_{k=-m}^{n-m} b_k \, s_{\tau-m-k} = d_{\tau-m} \backsimeq \delta_{\tau-m}, \tag{5.8}$$

and, relative to this new time origin, the output would become:

$$d_{-m} \, , \, d_{1-m} \, , \, \ldots d_{-1}, d_0, d_1, \, \ldots ,$$

$$\uparrow \qquad\qquad\qquad\qquad \uparrow$$

$$\tau = 0 \qquad\qquad\qquad \tau = m$$

and the filter response would be:

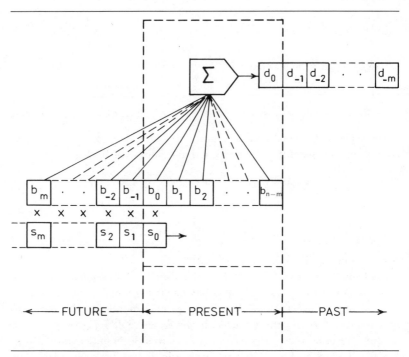

Figure 5.1. Convolution of a causal wavelet s_t *with a filter* b_t *to form a noncausal desired output* d_t. *The situation illustrated is at time* $t = 0$. *The filter* b_t *must be noncausal.*

$$b_{-m}, b_{1-m}, \ldots, b_{-1}, b_0, b_1, \ldots, b_{n-m}.$$

$$\uparrow \qquad\qquad\qquad \uparrow \qquad\quad \uparrow$$

$$\tau = 0 \qquad\qquad \tau = m \qquad \tau = n$$

In words, the result would be exactly the same, except that both the filter and its output would be causal. The price we pay for this is that the output is now an approximation to $\delta_{\tau-m}$, which is an impulse at time $\tau = m$. That is, relative to the new time origin, the impulse is delayed by m time units.

We conclude that it is possible to find a causal approximation to the inverse of a mixed-phase wavelet if we allow the output of the approximate inverse filter to be *delayed*. In practice, since we are dealing with recorded data, this delay is hardly an inconvenience.

It is now a straightforward matter to set up the design of a Wiener filter to turn our known mixed-phase wavelet s_t into a spike, provided we delay the spike. Referring to figure 1.4 and equations (1.15), we choose our desired output z_t to be the delayed spike:

$$z_t = 0, 0, \ldots, 0, 1, 0, \ldots = \delta_{t-m}. \tag{5.9}$$

$$\uparrow \qquad\quad \uparrow$$

$$t = 0 \qquad t = m$$

The input is the known wavelet s_t:

$$x_t = s_t = s_0, s_1, s_2, \ldots, \tag{5.10}$$

and the normal equations are:

$$\sum_{k=0}^{n} b_k \, \phi_{ss} \, (k - \tau) = \phi_{zs}(\tau), \qquad \tau = 0, 1, \ldots, n, \tag{5.11}$$

in which b_k is the filter, and the autocorrelation function $\phi_{ss}(\tau)$ and the cross-correlation function $\phi_{zs}(\tau)$ are given by:

$$\phi_{ss}(\tau) = \sum_{t=\tau}^{p} s_t \, s_{t-\tau}, \tag{5.12}$$

and

$$\phi_{zs}(\tau) = \sum_{t=\tau}^{\infty} z_t \, s_{t-\tau}. \tag{5.13}$$

We see immediately that these right-hand-side cross-correlation coefficients are simply:

$$\phi_{zs}(0) = s_m.$$

$$\phi_{zs}(1) = s_{m-1}.$$

$$\vdots \qquad \vdots$$

$$\phi_{zs}(m) = s_0.$$

$$\phi_{zs}(m+1) = 0.$$

$$\vdots \qquad \vdots$$

$$\phi_{zs}(n) = 0. \tag{5.14}$$

Thus the normal equations (5.6) are:

$$
\begin{bmatrix}
\phi_{ss}(0) & \phi_{ss}(1) \dots \phi_{ss}(n) \\
\phi_{ss}(1) & \phi_{ss}(0) \dots \phi_{ss}(n-1) \\
\vdots & \vdots \qquad \vdots \\
\phi_{ss}(m) & \phi_{ss}(m-1) \dots \phi_{ss}(n-m) \\
\phi_{ss}(m+1) & \phi_{ss}(m) \dots \phi_{ss}(n-m-1) \\
\vdots & \vdots \qquad \vdots \\
\phi_{ss}(n) & \phi_{ss}(n-1) \dots \phi_{ss}(0)
\end{bmatrix}
\begin{bmatrix}
b_0 \\ b_1 \\ \vdots \\ b_m \\ b_{m+1} \\ \vdots \\ b_n
\end{bmatrix}
=
\begin{bmatrix}
s_m \\ s_{m-1} \\ \vdots \\ s_0 \\ 0 \\ \vdots \\ 0
\end{bmatrix}
\tag{5.15}
$$

Thus, if the wavelet is mixed-phase, it is possible to find an approximate Wiener inverse provided the output can be delayed. We see that it is then essential to know both the autocorrelation $\phi_{ss}(\tau)$ of the wavelet and the first $m+1$ coefficients of the wavelet s_t. We should note that the weapon of delay is also essential if we wish to find an approximate causal inverse of a maximum-phase wavelet. This delay is necessary only to compensate for the anticausal part of the inverse; a minimum-phase wavelet has a causal inverse, and the delay is unnecessary.

Normally it is not worth troubling to find out whether the wavelet is minimum-phase, maximum-phase, or mixed-phase. Provided the desired output is a delayed spike, an approximate inverse can be found using equations (5.1). Since the inverse is only approximate, there will be an error consisting of nonzero values both before and after the main peak at time $t = m$. This is a kind of noise.

In applying such a filter to data, we must remember that it is deliberately designed to delay the output, relative to the original time origin. That is,

$$s_t * b_t = d_t \simeq \delta_{t-m}. \tag{5.16}$$

If we have a noise-free seismogram x_t that is the convolution of our known wavelet s_t with the discretely sampled unknown impulse response of the earth g_t:

$$x_t = s_t * g_t, \tag{5.17}$$

then the application of our spiking filter b_t to the seismogram will give the result:

$$
\begin{aligned}
x_t * b_t &= s_t * g_t * b_t \\
&= (s_t * b_t) * g_t \\
&= d_t * g_t \\
&\simeq \delta_{t-m} * g_t = g_{t-m}.
\end{aligned} \tag{5.18}
$$

The result is thus an approximate delayed version of the impulse response of the earth. In order to preserve the timing of the data after application of this filter, we need to shift the data back m samples, thus losing the first m samples of the output. These m samples will be precursor noise introduced by the errors in the filter design. Every event in the recovered response is preceded by some precursor noise and followed by postcursor noise. In general these two types of noise will overlap: the postcursor noise of an early event will occur simultaneously with precursor noise from a later event. This deconvolution noise arises from the impossibility of designing an exact inverse of a finite-length wavelet.

5.2. Spiking Deconvolution of a Known Wavelet in the Frequency Domain

Some insight into what we are trying to do with these time-domain filters can be obtained by looking at the operation in the frequency domain. We recall from chapter 3, section 3.3, that the Fourier transform of a wavelet is related to the z-transform by the unit circle:

$$z = e^{-2\pi i f \Delta t},$$

where f is the frequency and Δt is the sampling interval. From equation (5.4) we can see immediately that the Fourier transform $A(f)$ of the inverse of the wavelet is related to the Fourier transform $S(f)$ of the wavelet by:

$$A(f) = \frac{1}{S(f)}, \tag{5.19}$$

where both $A(f)$ and $S(f)$ are complex functions of frequency. Expressing this equation in terms of amplitude and phase, we have:

$$|A(f)| \, e^{i \phi_a(f)} = \frac{1}{|S(f)| \, e^{i \phi_s(f)}}, \tag{5.20}$$

from which we may deduce:

$$|A(f)| = \frac{1}{|S(f)|},$$ (5.21)

and

$$\phi_a(f) = -\phi_s(f).$$ (5.22)

Of particular interest is equation (5.21), which states that the amplitude spectrum of the inverse filter is the inverse of the amplitude spectrum of the wavelet. Very often real wavelets have spectra that are deficient in certain frequencies. This is particularly true of marine wavelets, which suffer from the effect of the sea-surface reflection, which introduces notches in the spectrum. At these points in the spectrum $|S(f)|$ is very small, and $|A(f)|$ must be very large.

Most seismic data has been recorded digitally, with the analog-to-digital conversion taking place after the signal has been filtered with a low-pass antialias filter. This typically has a corner frequency at half the Nyquist frequency and cuts off at about 72 dB per octave. Thus any signal (and noise) above the Nyquist frequency has been attenuated by more than 72 dB, and the bandwidth of the remaining data extends no further than the Nyquist frequency where the data level is effectively zero.

A noise-free signal spectrum might look as shown in figure 5.2, where we also show the spectrum of the inverse filter. The enormous amplitudes of the inverse spectrum at frequencies where the signal spectrum is small are a cause for worry: in the time domain the filter will oscillate uncontrollably at these frequencies. In order to control this violent oscillation, prior to computing the inverse some energy must be added to the signal at frequencies where $|S(f)|$ is small.

In computing the approximate inverse b_t, we must expect similar problems. In fact these do occur unless precautions are taken. The solution is elegant and is now well known.

Instead of selectively adding energy at only those frequencies where it is needed, the energy is added at all frequencies in the same amount. That is, a small positive constant is added to the power spectrum (see Appendix) of the wavelet. It is not done in the frequency domain, however; it is done directly in the normal equations. The power spectrum of s_t is

$$\Phi_{ss}(f) = |S(f)|^2.$$ (5.23)

Adding a small positive constant W to this spectrum, we form the modified spectrum $\Phi_{ss}(f) + W$, where W is known as a *white-noise spectrum*. The Fourier transform of $\Phi_{ss}(f)$ is the autocorrelation function $\phi_{ss}(\tau)$ (see Appendix). The Fourier transform of a white-noise power spectrum (or constant in the frequency domain) is a spike autocorrelation function, $w \cdot \delta(\tau)$, where w is a constant, shown in figure 4.3 as σ^2.

Thus the effect of white noise is simply to add a constant to the zero-lag

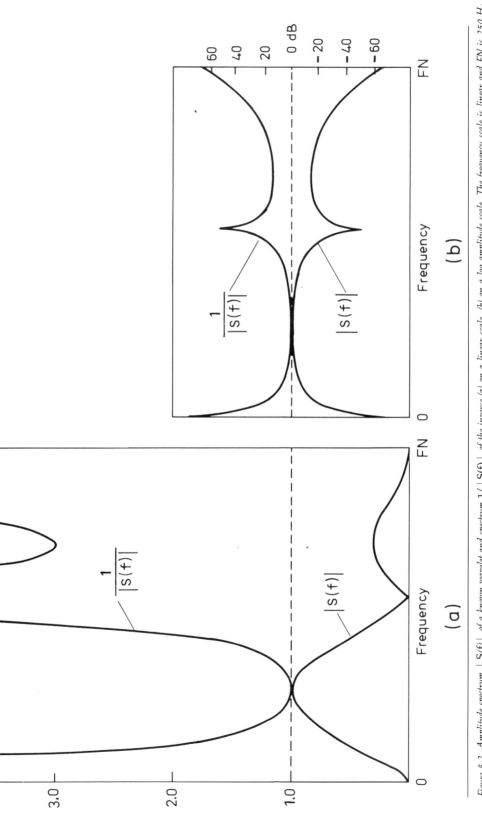

Figure 5.2. Amplitude spectrum $|S(f)|$ of a known wavelet and spectrum $1/|S(f)|$ of the inverse (a) on a linear scale, (b) on a log amplitude scale. The frequency scale is linear and FN is 250 H_z.

autocorrelation coefficient of the signal. Therefore, in order to stabilize the solution of the normal equations, a small positive constant is always added to the zero-lag autocorrelation coefficient. This is often called "adding white noise."

5.3. The Problem of Noise

In practice we never deal with noise-free signals. (Even in synthetic studies there is computer round-off error, which is a kind of noise.) Usually there is added noise, and the seismic signal x_t can be described by the normal convolution plus noise n_t:

$$x_t = s_t * g_t + n_t$$
$$= \sum_{k=0}^{p} s_k g_{t-k} + n_t. \tag{5.24}$$

Let us suppose that we have designed an approximate inverse spiking filter b_t according to the method described in section 5.1. How will this filter perform on data contaminated by noise? Applying the filter to the data x_t, we have the output y_t:

$$y_t = x_t * b_t$$
$$= s_t * g_t * b_t + n_t * b_t, \tag{5.25}$$

and we see that we have the extra term formed by the convolution of the filter b_t with the noise n_t. In the frequency- domain equation (5.25) becomes:

$$Y(f) = S(f) \cdot G(f) \cdot B(f) + N(f) \cdot B(f), \tag{5.26}$$

where we have used the notation of section (5.2) to express the Fourier transform relations.

In equation (5.26), $B(f)$ is approximately the inverse of $S(f)$:

$$B(f) \approx \frac{1}{S(f)}. \tag{5.27}$$

In particular, the amplitude spectrum of the filter is approximately the inverse of the amplitude spectrum of the wavelet:

$$|B(f)| \approx \frac{1}{|S(f)|}, \tag{5.28}$$

which means that the filter has a large gain at frequencies where the wavelet has low energy, and low gain at frequencies where the wavelet has high energy. The effect of the filter on the noise is to blow up the noise at frequencies where the signal is weak, as illustrated in figure 5.3.

The filter b_t has been designed without regard to the noise n_t. It is possible that we could design a better filter by taking the noise into account. Ideally

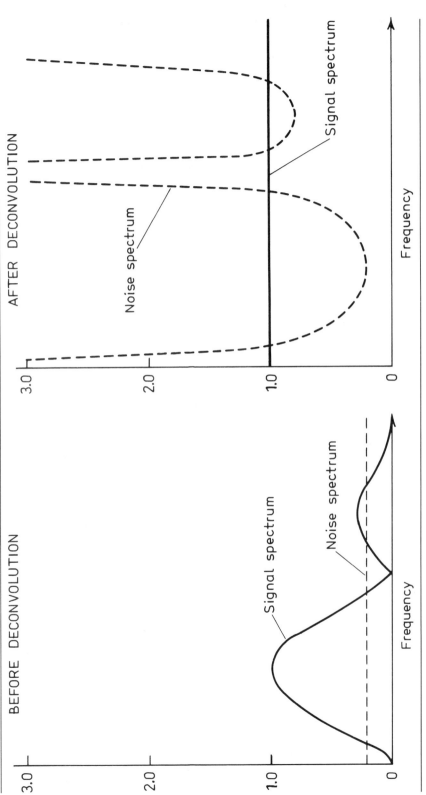

Figure 5.3. The effect of spiking signature deconvolution on the spectral amplitude of the noise.

we should like the filter output to be g_t when the input is x_t. We know, however, that for the noise-free case the best we can do, if our wavelet is not minimum-phase, is to design a filter whose output approximates a *delayed version* of g_t. That is, the best we can aim for is an output that is an approximation to

$$g_{t-m} = \delta_{t-m} * g_t. \tag{5.29}$$

When we have noise, we can try to design our least-squares filter b_t such that the desired output is g_{t-m} when the input is x_t The normal equations are then:

$$\sum_{k=0}^{n} b_k \, \phi_{xx}(k - \tau) = \phi_{gx}(\tau - m), \qquad \tau = 0, 1, \ldots, n, \tag{5.30}$$

where

$$\phi_{gx}(\tau - m) = \sum_{s=\tau-m}^{\infty} g_s \, x_{s-\tau+m}$$

$$= \sum_{t=\tau}^{\infty} g_{t-m} x_{t-\tau}. \tag{5.31}$$

In equations (5.30) x_t is known and therefore $\phi_{xx}(k - \tau)$ can be estimated as discussed in chapter 4. The desired output g_t is not known. Therefore, the equations cannot be solved.

An approximate solution to this problem has been proposed by Robinson and Treitel (1967) and is reproduced in their book (1980, pp. 155–158). They assume that the noise n_t is a stationary nonwhite sequence that can be regarded as the convolution of a wavelet c_t with a stationary white-noise sequence q_t:

$$n_t = c_t * q_t. \tag{5.32}$$

They also assume that the impulse response of the earth g_t is white, random, and stationary with variance σ^2 and is uncorrelated with q_t. After some algebra, Robinson and Treitel manipulate the normal equations (5.30) into the form:

$$\sum_{k=0}^{n} b_k \, [\sigma^2 \, \phi_{ss} (k - \tau) + \phi_{nn} (k - \tau)] = \sigma^2 s_{m-\tau}$$

$$\text{for } \tau = 0, 1, \ldots, n, \tag{5.33}$$

and we see that it is not necessary to know g_t. It is only necessary to have an estimate of the autocorrelation of the uncorrelated noise $\phi_{nn}(\tau)$ plus an estimate of σ^2. Since the wavelet s_t is known, the equations can be solved for the filter coefficients b_k.

It is worth considering whether the assumptions that must be made to reach the solution (5.33) are likely to be valid. If the uncorrelated noise n_t is stationary and additive, as we have assumed, it is something that is going on whether or not we do our seismic experiment. It is *uncorrelated* with our experiment. It exists independent of the experiment; and when we record our data x_t, they are recorded simultaneously with our signal $s_t * g_t$.

We must also assume that g_t is stationary. As we have discussed in earlier chapters, particularly chapter 2, g_t cannot be stationary if the point-source convolutional model is valid. In order to force g_t to be stationary, we have to apply a spherical-divergence correction to the recorded data x_t. Apart from violating the convolutional model, as discussed in chapter 2, this spherical-divergence correction must force the noise to be nonstationary.

In other words, we cannot assume that both g_t and n_t are stationary simultaneously. Therefore, we must question whether equations (5.33) give the best design of the Wiener spiking filter in the presence of noise.

Noise is a real problem, of course. Because the spiking filter blows the noise up at frequencies where the signal is weak, as shown in figure 5.3, the data must be filtered to reduce the noise level. That is, outside the band where the signal-to-noise ratio is good, the noise is amplified too much by spiking deconvolution and must be suppressed by band-pass filtering, in order that the signal can be seen (figure 5.4).

The two operations of spiking filtering and band-pass filtering can be combined into one operation in which the desired output of the Wiener filter is not a spike but a wavelet with the desired bandwidth. What is the best wavelet?

5.4. The Advantage of the Zero-Phase Wavelet

It was proved by Berkhout (1974) that the shortest wavelet having a given bandwidth is zero-phase. The relationship between a zero-phase wavelet and any other wavelet with the same amplitude spectrum is as follows.

Let $s(t)$ be a *real* wavelet with Fourier transform

$$S(f) = \int_{\infty}^{\infty} s(t) \, e^{-2\pi i f t} \, dt$$
$$= |S(f)| \, e^{i\phi_s(f)}, \tag{5.34}$$

where $|S(f)|$ is the amplitude spectrum and $\phi_s(f)$ is the phase spectrum. When $s(t)$ is real, $S(f)$ exhibits hermitian or complex-conjugate symmetry

$$S(-f) = S^*(f)$$
$$= |S(f)| \, e^{-i\phi_s(f)}. \tag{5.35}$$

We can construct a wavelet from this Fourier transform, which has the same amplitude spectrum but no phase spectrum. That is,

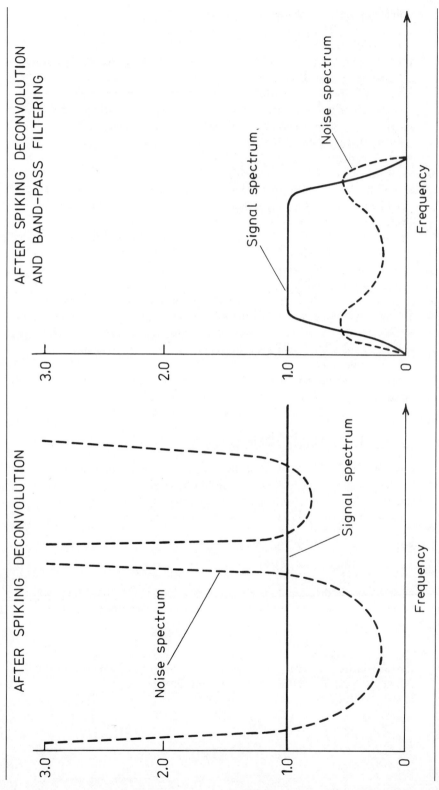

Figure 5.4. Band-pass filtering is often necessary after spiking deconvolution to prevent the signal from being drowned in the noise.

$$S(-f) = S(f) = |S(f)|. \tag{5.36}$$

The wavelet that has this Fourier transform is denoted by $s^0(t)$, which is called the *zero-phase correspondent* of the wavelet $s(t)$ with the given amplitude spectrum. The wavelet $s^0(t)$ is related to the amplitude spectrum $|S(f)|$ by the inverse Fourier transform relation

$$
\begin{aligned}
s^0(t) &= \int_{-\infty}^{\infty} |S(f)| \, e^{2\pi i f t} df \\
&= \int_{-\infty}^{\infty} |S(-f)| \, e^{-2\pi i f t} df \\
&= s^0(-t).
\end{aligned} \tag{5.37}
$$

Thus this real wavelet is *symmetric* and therefore *noncausal*. Any causal wavelet cannot be zero-phase.

In seismic data our wavelets are always real and causal. Therefore, we always have the possibility of filtering the data to turn our known wavelet into its zero-phase correspondent. The reason we would want to do this is that the zero-phase correspondent has the minimum length of any wavelet in the suite that shares a given amplitude spectrum; therefore, we will improve the resolution of the data if we make the wavelet zero-phase—even if we do not alter the amplitude spectrum. In practice, we will probably want to broaden the amplitude spectrum somewhat—within the limits of the bandwidth defined by good signal-to-noise ratio.

Of course, we cannot deal with noncausal wavelets, as discussed earlier in this chapter. We can, however, deal with delayed versions of noncausal wavelets. If s_t is our known discretely sampled real seismic wavelet, then s^0_t is its noncausal zero-phase correspondent and s^0_{t-m} is a delayed version of s^0_t. By choosing m large enough, we can always make s^0_{t-m} causal.

Typically we have a known mixed-phase wavelet s_t with amplitude spectrum $|S(f)|$. We wish to improve the resolution of the data (1) by broadening the spectrum within the limits imposed by the bandwidth where the signal-to-noise ratio is good and (2) by ensuring that the final wavelet is a delayed version of the zero-phase correspondent with the desired bandwidth. If $|D(f)|$ is the desired amplitude spectrum of the final wavelet, with zero-phase correspondent d^0_t, then the desired output of our Wiener filter would be the causal wavelet d^0_{t-m}. The Wiener filter would then be designed to shape the known wavelet s_t into the desired wavelet d^0_{t-m}. The operations are illustrated in figure 5.6. The result is equivalent to spiking deconvolution followed by band-pass filtering with a zero-phase band-pass filter.

5.5. Minimum-Phase Signature Deconvolution

As we have seen in chapter 4, it is quite common to attempt to remove multiples with prediction-error filtering, in which the predictable part must

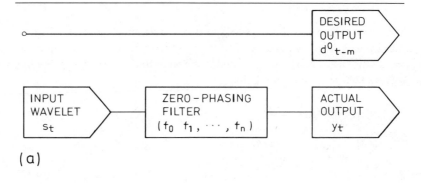

(a)

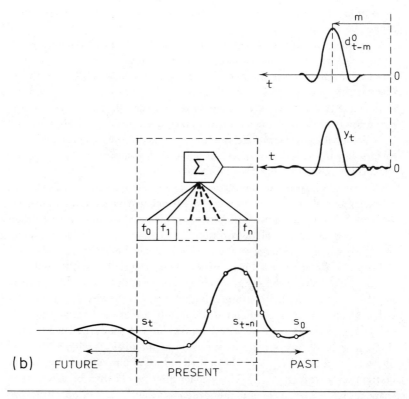

(b)

Figure 5.5. (a) Design elements of the zero-phasing Wiener filter. (b) The operation of the filter.

be minimum-phase. If the wavelet is not minimum-phase, prediction-error filtering may not work. If our known wavelet s_t is not minimum-phase, we can try to make it minimum-phase by filtering. That is, we can find a minimum-phase wavelet $d_{t\text{min}}$ that has the desired amplitude spectrum $|D(f)|$, and we find the Wiener filter f_t that shapes our known wavelet s_t into our best approximation of $d_{t\text{min}}$. Application of f_t to the data will then force the wavelet to be more nearly minimum-phase. This filtering step must be done before the application of any spherical-divergence correction, of course, as it must with any signature deconvolution.

In designing the Wiener filter f_t, it will probably be necessary to delay the desired wavelet $d_{t\text{min}}$ by m time units. The desired output of the Wiener filter should therefore be $d_{t-m\ \text{min}}$, and the normal equations are then:

$$\sum_{k=0}^{n} f_k\ \phi_{ss}\ (\tau - k) = \phi_{ds}\ (\tau - m), \quad \tau = 0, 1, \ldots, n, \tag{5.38}$$

where

$$\phi_{ds}(\tau - m) = \sum_{t=\tau}^{\infty} d_{t-m} s_{t-\tau}. \tag{5.39}$$

It is now quite common for initial signature deconvolution to be of this type. It permits predictive deconvolution to be done later; and if there is still any confidence that the phase of the wavelet has been preserved right through to the stack, then a final deconvolution to zero-phase is often done at the very end to obtain maximum resolution in the stacked data.

There is a problem with the minimum-phase deconvolution described here. If the wavelet s_t is mixed phase and we want to filter it to make it minimum-phase, we will have to use a noncausal filter. We cannot do this, and we have no alternative but to delay the desired minimum-phase output as described earlier. Because the filter will not be exact, there will be some precursor noise as discussed in section 5.1. That is, before the onset of our approximate minimum-phase wavelet, we will always have some precursor noise. Thus every minimum-phase wavelet overlaps the precursor noise of subsequent minimum-phase wavelets. To put it another way, we can try to make our wavelet minimum-phase, but we will be in error and will generate noise by doing so. This noise is not random; it is a function of our filter design method, and it will correlate from trace to trace and will most probably stack up.

However you look at it, deconvolution of the signature is a process that adds noise to the data.

6. EXTRACTION OF AN UNKNOWN
WAVELET

6.1. Introduction

In the one-dimensional convolutional model of the seismogram, we propose that each seismic trace consists of the convolution of a seismic wavelet $s(t)$ with the impulse response of the earth $g(t)$, plus some added noise $n(t)$. Thus,

$$x(t) = s(t) * g(t) + n(t). \tag{6.1}$$

The signal $x(t)$ is the time-varying pressure, particle velocity, or particle acceleration that we try to measure with our device—hydrophone, geophone, or accelerometer. Very often we do not have a point receiver: the outputs from several receivers spread over some finite area are usually summed to form a single signal, which is then passed through a filter network and recorded on magnetic tape. Thus the incident signal $x(t)$ is distorted to some extent by the array of measuring receivers and the recording system. We now ignore all this distortion; and we suppose that what we record on tape faithfully represents the incident signal $x(t)$ at one point in space. That is, we say we know $x(t)$, the left-hand side of equation (6.1).

On the right-hand side the quantity we wish to find is $g(t)$, which we want to estimate accurately in the presence of the noise $n(t)$, even when the signal $s(t)$ is unknown. That is, all three quantities on the right-hand side of equation (6.1) are unknown.

The purpose of this chapter is to state as simply as possible that any method that claims to solve equation (6.1) for $g(t)$, given only $x(t)$, must be regarded with great suspicion. Clearly, it is impossible to solve one equation with three unknowns. Nevertheless, some of the most interesting papers in the geophysical literature offer plausible solutions to this problem; the reader often has to work very hard to see where the restricting assumptions are made. In the remainder of this chapter a couple of well-known state-of-the-art wavelet-extraction techniques are discussed, with particular emphasis on the assumptions on which they are based.

6.2. Standard Prediction-Error Deconvolution Extraction

The most well known method of wavelet extraction is based on the minimum-phase property of the one-step-ahead least-squares prediction-error filter. It has already been discussed in sections 3.7 and 4.6.

The basic steps in the extraction are as follows:

1. Apply a spherical-divergence correction, as described in section 2.7, in an attempt to force the data to be stationary. The price paid for this stationarity is *uneven distortion of the wavelet* and violation of the convolutional model.
2. Find the autocorrelation of the seismic trace over some desired window. If the *impulse response of the earth is assumed to be white, random, and stationary,* then the autocorrelation function is basically the same as the autocorrelation of the wavelet.
3. Find the least-squares one-step-ahead prediction-error filter, as described in section 3.3. For this calculation, only the autocorrelation needs to be known. This prediction-error filter is the least-squares inverse of the minimum-phase wavelet that has the computed autocorrelation.
4. Find the inverse of the filter. This can be done by least squares again. As described in sections 3.7 and 4.6, this inverse is the least-squares approximation of the minimum-phase wavelet that has the computed autocorrelation function. In other words, *the wavelet is assumed to be minimum-phase.*

The errors that are introduced by these processing steps have already been discussed in section 4.6 using an example. The example does not deal with the problem of added noise or the problem that occurs if the impulse response of the earth is not white, random, and stationary. If there is added noise, the estimate of the autocorrelation function will be less accurate. If the earth impulse response is not stationary, the estimate of the autocorrelation function will also be in error. If the earth impulse response is not white, the estimated autocorrelation will not be purely a scaled version of the autocorrelation of the wavelet; it will also include contributions at nonzero lags from the geology. The wavelet that is estimated will include some of these geological effects, and deconvolution to remove the wavelet will be partly successful on the inaccurately estimated wavelet; it will also tend to remove some of the primary reflections. Unless we happen to *know* that the impulse response of the earth is white, there is always a risk that we will be attenuating primary reflections while we mistakenly believe we are simply removing the effect of the wavelet.

The effect of the spherical divergence should not be ignored. We are forced to apply a spherical-divergence correction in order to average out the effects of the geology on the estimate of the autocorrelation of the wavelet. This correction forces an error, as we saw in the example in section 4.6. Therefore, even if *all* our assumptions about the geology and the wavelet are correct, we *still* cannot estimate the wavelet accurately with this method.

6.3. Exponential Tapering Method

From Robinson's energy-delay theorem we know that, of all the wavelets that share a given autocorrelation function, the minimum-phase wavelet has its energy concentrated as close to the beginning as possible. The energy in a non-minimum-phase wavelet is delayed relative to the energy in the corresponding minimum-phase wavelet. It would be possible to force a non-minimum-phase wavelet to be minimum-phase (but with a different autocorrelation function) if the front of the wavelet were emphasized relative to the tail by the application of some kind of taper, as shown in figure 6.1. If the autocorrelation of this tapered minimum-phase wavelet could be estimated, the wavelet could be found and the original wavelet could then be recovered by inverse tapering. This idea is sometimes used in a state-of-the-art wavelet-extraction scheme that is included in some commercially available seismic data-processing packages.

Consider equation (6.1), in which the noise term $n(t)$ is neglected:

$$x(t) = s(t) * g(t)$$
$$= x_0, x_1, x_2, \ldots, x_N. \tag{6.2}$$

The z-transform in this equation is:

$$X(z) = S(z) \cdot G(z)$$
$$= (s_0 + s_1 z + \ldots + s_n z^n) \cdot (g_0 + g_1 z + \ldots + g_{N-n} z^{N-n})$$
$$= x_0 + x_1 z + x_2 z^2 + \ldots + x_N z^N. \tag{6.3}$$

We may factorize this z-transform into N roots, of which n are the roots of $S(z)$ and $N - n$ are the roots of $G(z)$:

$$X(z) = x_N \prod_{i=1}^{N} (z - z_i)$$

$$= [s_n \prod_{j=1}^{n} (z - z_j)] \cdot [g_{N-n} \prod_{k=1}^{N-n} (z - z_k)]. \tag{6.4}$$

Consider now the application of an exponential taper $e^{-\alpha t}$ to the seismogram $x(t)$, in which α is a small positive number and $e^{-\alpha}$ is therefore slightly less than one. We have:

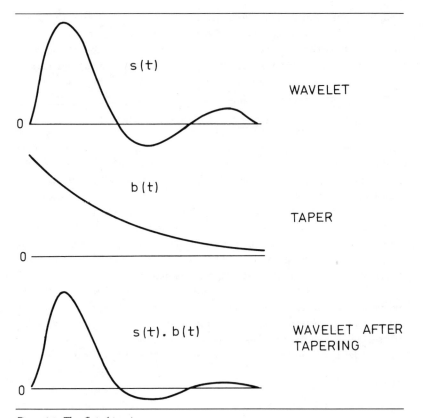

s(t)

b(t)

s(t).b(t)

WAVELET

TAPER

WAVELET AFTER
TAPERING

Figure 6.1. The effect of tapering.

$$\hat{x}\,(t) = e^{-\alpha t}\,x(t)$$
$$= x_0,\, e^{-\alpha}x_1,\, e^{-2\alpha}x_2,\, \ldots,\, e^{-N\alpha}x_N. \tag{6.5}$$

The z-transform of the tapered seismogram $\hat{x}\,(t)$ is:

$$\hat{x}\,(z) = x_0 + e^{-\alpha}x_1 z + e^{-2\alpha}x_2 z^2 + \ldots + e^{-N\alpha}x_N z^N$$
$$= x_0 + x_1(e^{-\alpha}z) + x_2(e^{-\alpha}z)^2 + \ldots + x_N(e^{-\alpha}z)^N. \tag{6.6}$$

Comparing equations (6.3) and (6.6), we see that wherever we have a z on the right-hand side of equation (6.3), we have $(e^{-\alpha}z)$ on the right-hand side of equation (6.6). Otherwise the equations are identical. We may therefore, by inspection, factorize $\hat{x}\,(z)$ into its N roots:

$$\hat{r}\,X(z) = x_N \prod_{i=1}^{N} (e^{-\alpha}z - z_i)$$
$$= [s_n \prod_{j=1}^{n} (e^{-\alpha}z - z_j)] \cdot [g_{N-n} \prod_{k=1}^{N-n} (e^{-\alpha}z - z_k)], \tag{6.7}$$

where the z_j are the roots of $S(z)$ and the z_k are the roots of $G(z)$.

From equation (6.4), we see that $X(z)$ has the N roots

$$z = z_i, \quad i = 1, 2, \ldots, N, \tag{6.8}$$

whereas, from equation (6.7), we see that $\hat{x}\,(z)$ has the N roots:

$$z = e^{\alpha}z_i, \quad i = 1, 2, \ldots, N, \tag{6.9}$$

in which e^{α} is a positive number slightly greater than one. That is, the exponential taper causes the roots of the z-transform to be shifted outward slightly, by a factor e^{α}.

A non-minimum-phase wavelet has at least one of its roots inside the unit circle. If the wavelet is tapered with an exponential taper, the roots will all move outward. If α is chosen sufficiently large, all the roots can be pushed outside the unit circle; that is, the wavelet becomes minimum-phase. We see that if the wavelet is convolved with another function, as in equation 6.2, the application of the taper has the same effect on all the roots, regardless of the convolution. Therefore, the seismogram can be tapered exponentially and the wavelet can be forced to be minimum-phase.

If we can estimate the autocorrelation of the wavelet, we can find the minimum-phase wavelet and simply apply the inverse exponential $e^{\alpha t}$ to recover the original wavelet.

The problem is: How do we estimate the autocorrelation of the wavelet? This is a nontrivial problem. Even before the exponential taper, the data were nonstationary. Normally, as we have seen, some sort of spherical-divergence correction must be applied to force the data to be approximately stationary so that the autocorrelation can be estimated. The spherical-divergence correction increases the amplitude of the tail of the seismogram. The exponential taper decreases the amplitude of the tail of the seismogram. It does exactly the opposite of what is required to make the seismogram stationary: it makes it even less stationary (see figure 6.2).

As a corollary we should note that the normal spherical-divergence correction, by amplifying the tail of the seismogram and therefore the tail of the wavelet, is undermining the normal assumption that the wavelet is minimum-phase.

Suppose we are convinced that the wavelet is not minimum-phase and are attracted by the possibilities of this exponential tapering technique. How do we estimate the autocorrelation of this tapered wavelet, which is now convolved with an extremely nonstationary sequence?

We rely on averaging. In the state-of-the-art packages that use this technique, the autocorrelation of the tapered seismogram is estimated for several windows and on many traces within the shot record. Thus for each shot record a number of autocorrelations are made. These different autocorrelations are then simply added together. The resulting summed autocorrelation is then regarded as the autocorrelation of the tapered wavelet, on the dubious ground that all the contributions of the nonstationary geology will have averaged out.

6.4. Sea-Bottom Reflection Method

Sometimes in marine seismic surveying there are not significant reflectors immediately beneath the sea floor. The first *significant* reflection after the sea-bottom reflection may not arrive until, say, $m + 1$ samples after the arrival of the sea-bottom reflection. Thus, referring to our convolutional model, section 2.3, our discrete reflections are g_1, g_2, g_3, and so on, and we identify g_1 as the sea-bottom reflection arriving at a time t_1, and g_2 as the first significant reflection beneath the sea floor, arriving at a time

$$t_2 = t_1 + (m + 1)\Delta t, \tag{6.10}$$

where Δt is the sample interval. The reflections g_2, g_3, and so on arrive at later and later times.

If this situation occurs, we may identify the first $m + 1$ samples of the reflection sequence as an uncontaminated version of the $m + 1$ samples of the returning seismic wavelet s_t, scaled by the amplitude of the sea-bottom reflection g_1. That is, the first $m + 1$ samples will be

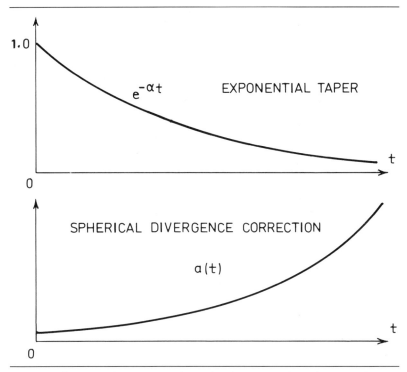

Figure 6.2. The exponential taper can force the wavelet to be minimum-phase. The spherical divergence correction forces the data to be more nearly stationary. The simultaneous assumptions of a minimum-phase wavelet and a stationary seismogram are often inconsistent.

$$s_0 g_1, \; s_1 g_1, \; s_2 g_1 \; \cdots \; , \; s_m g_1,$$

$$\uparrow \qquad\qquad\qquad\qquad \uparrow$$

$$t = t_1 \qquad\qquad\qquad\quad t = t_1 + m\,\Delta t$$

or $g_1 \, (s_0, s_1, s_2, \ldots , s_m)$.

If the autocorrelation of s_t can be estimated approximately from the data as described in chapter 4, then a non-minimum-phase filter b_t may be found from equations (5.15); this filter is a least-squares approximation to the inverse of s_t, as described in section 5.1. An approximation to the full wavelet s_t may be found by finding the inverse of b_t. If it is suspected that s_t is not minimum-phase, then the equations that must be solved to find s_t are:

$$
\begin{bmatrix}
\phi_{bb}(0) & \phi_{bb}(1) & \cdots & \phi_{bb}(n) \\
\phi_{bb}(1) & \phi_{bb}(0) & \cdots & \phi_{bb}(n-1) \\
\vdots & \vdots & & \vdots \\
\phi_{bb}(m) & \phi_{bb}(m-1) & \cdots & \phi_{bb}(n-m) \\
\phi_{bb}(m+1) & \phi_{bb}(m) & \cdots & \phi_{bb}(n-m-1) \\
\vdots & \vdots & & \vdots \\
\phi_{bb}(n) & \phi_{bb}(n-1) & \cdots & \phi_{bb}(0)
\end{bmatrix}
\begin{bmatrix}
s_0 \\ s_1 \\ \vdots \\ s_m \\ s_{m+1} \\ \vdots \\ s_n
\end{bmatrix}
=
\begin{bmatrix}
b_m \\ b_{m-1} \\ \vdots \\ b_0 \\ 0 \\ \vdots \\ 0
\end{bmatrix}
. \quad (6.11)
$$

For this method to be successful, it is necessary to have an estimate of the autocorrelation of the wavelet, and to *know* that the first $m + 1$ samples of the returning seismic reflection data are purely associated with the sea-bottom reflection, and are not contaminated by any deeper reflections.

This method differs from the two methods discussed in sections 6.2 and 6.3 in that a direct measurement of *part* of the wavelet is made. There is uncertainty, however, about how much of the wavelet is measured without contamination from deeper reflections.

6.5. Objections to These and Other Statistical Methods

In addition to the methods described earlier, which attempt to extract the seismic wavelet from the data, there are many others, including Kalman filtering, homomorphic deconvolution, minimum-entropy deconvolution, parsimonious deconvolution, and so on. A number of these methods are described in Webster (1978) and in Berkhout and Zaanen (1976). In all of them the wavelet is not known from direct measurements: it is extracted directly from the data.

Each of these methods starts with a model of the data corresponding to equation (6.1) and follows with assumptions about the statistical properties of the earth impulse response $g(t)$ and the noise $n(t)$. Very often assumptions also have to be made about the properties of the wavelet $s(t)$. In general, with the exception of homomorphic deconvolution, these methods often need to assume something about the stationarity of the data and happily invoke the plane-wave convolutional model of the seismogram, whether it applies or not. As we saw in section 5.3, if the point-source convolutional model applies, it is not possible to assume that both $g(t)$ and $n(t)$ are stationary. If $n(t)$ is stationary on the recorded data, $g(t)$ will not be stationary because of spherical divergence. If a spherical-divergence correction is applied to the data $x(t)$ in an attempt to force $g(t)$ to be stationary, it will necessarily increase the amplitude of the tail of the noise $n(t)$, thus making it nonstationary. As we saw in chapter 4, long windows or many short windows averaged are necessary to make good statistical estimates; the longer the window, the greater the distortion of the spherical-divergence correction.

These statistical methods are distinguished from each other by the assumptions that are made about the statistical properties of the data. Each method assumes a different statistical model for the data, and for any given data set each method will necessarily extract a different wavelet. Thus we will have as many different wavelets as there are different methods.

Of course, if our convolutional model (eq. 6.1) is right, there is only one true wavelet present in the data. How can we determine which, if any, of these statistical methods is able to extract the true wavelet? One possibility is to test the method against a known wavelet. To some extent we have done this in the synthetic example of section 4.6. In this case the statistical assumptions about $g(t)$ were correct; also, there was no noise $n(t)$. The assumption that the wavelet was minimum-phase was incorrect, however. Figure 4.11 shows the true wavelet and the minimum-phase correspondent of that wavelet. Even in these ideal circumstances, the need for the spherical-divergence correction introduces distortion such that the extracted wavelet is not even close to the minimum-phase correspondent of the true wavelet, as shown in figure 4.13.

This synthetic test shows the magnitude of the error that can be made by the need for the spherical-divergence correction. It also shows the importance of knowing the phase of the wavelet. With real data the situation is much worse. First, we have noise $n(t)$, which is unknown; second, we have an earth impulse response $g(t)$ that may or may not have the statistical properties that are attributed to it. We could test the methods by comparing the wavelets extracted with a measured wavelet. In the case of marine seismic exploration, "the wavelet" can usually be measured if the water is deep enough. The statistical extraction methods are applied to data shot in shallow water—the usual case—where the water is not deep enough to permit the measurement of the wavelet to be made properly. To my knowledge, there is no published work comparing wavelets extracted from marine seismic data with those measured in deep water. If such comparisons were made the wavelets might be different. The differences could be attributed to (1) inaccuracies in the wavelet extraction method; or (2) propagation effects which are observed only in the extracted wavelet; or (3) some combination of these two factors. It is not possible to separate the suspected propagation effects from inaccuracies in the wavelet extraction method.

In the absence of any reliable test, new statistical deconvolution methods continue to be published.

One must consider seriously how much validity any such test would have. It could be, for example, that the test is conducted in a place where the statistics of the geology and the noise happen to be close to what is assumed by one of these wavelet-extraction methods. The wavelet extracted by this particular method will then be similar to the measured wavelet. The wavelets extracted by all the other methods will be less similar. One cannot conclude from this that the method that was successful in the place where the test was conducted will be successful in other places, where the statistics of the geology are unknown and may be different. In these places, other wavelet-extraction techniques may be more successful.

If we know the wavelet, we can test the wavelet-extraction technique. The point is not to extract the wavelet, however; the point is to extract the earth impulse response $g(t)$. If we already know the wavelet, we have no need of a wavelet-extraction technique. We may try to extract $g(t)$ by the methods described in chapter 5. If we do not know the wavelet, we can use any of the statistical wavelet-extraction techniques to find "the wavelet" and then use this "known" wavelet to find the impulse response of the earth $g(t)$. Each method will lead to a different wavelet and a different $g(t)$. How do we know which method is right?

We could compare the extracted $g(t)$ with a synthetic impulse response calculated from a well log, bearing in mind the problems we are likely to face (see section 1.6). One of the wavelet-extraction methods will give a better fit of the extracted $g(t)$ to the synthetic seismogram than any of the other methods, though whether the fit is "good" or not may be difficult to say. Once again, however, we do not know that this method will apply to all the places where we do not have wells to test the validity of the method. Equally, it follows that the other methods may be more valid in other places.

Therefore, if we cannot compare the results with a measured wavelet or a synthetic seismogram calculated from well logs, we have no idea how reliable any of these methods is likely to be. Furthermore, there is no test we can apply that can refute any of these methods. There is always a chance that any of these methods might work.

The point of seismic surveying is to find the places at which it pays to drill. We need to apply deconvolution to sharpen up the image so that we may accurately assess in which places drilling is most likely to be successful. In such places we need to be able to rely on our deconvolution. It is in exactly these places, however, that we cannot depend on our statistical wavelet-extraction methods.

In his book *Conjectures and Refutations* (1972), Sir Karl Popper makes a demarcation between two kinds of theories: scientific theories and nonscientific theories. Scientific theories are framed in such a way that in principle they stand at risk of being refuted by a test. A nonscientific theory is framed in such a way that in principle it cannot be refuted by a test. We have just seen that neither of the two kinds of tests we have been able to devise (comparing the extracted wavelet with a measured wavelet, and comparing the extracted $g(t)$ with a well-synthetic seismogram) can refute the theories on which the wavelet-extraction methods are based. These methods would therefore be regarded as nonscientific on Popper's demarcation criterion, as pointed out in detail in Ziolkowski (1982).

The fact is that when we have one equation (eq. 6.1) and three unknowns, $s(t)$, $g(t)$, and $n(t)$, we cannot expect to determine any of the unknowns unless we supply outside information. The statistical wavelet-extraction techniques supply guesses instead of information. The guesses can be quite sophisticated, but they are still guesses; that is why we must regard all these techniques with deep suspicion.

7. PHASE ERROR AND THE

INTERPRETATION OF LITHOLOGY

Throughout this book emphasis has been placed on the assumptions that underlie both the convolutional model of the seismogram and the methods that are used to deconvolve seismic data. Much of the argument has been critical of current methods, especially where no measurement of the wavelet is available.

At this point the reader may have some sympathy with some of the criticism, but may wonder whether or not it adds up to something significant. For example, we know that the spherical-divergence correction distorts the wavelet unevenly down the seismogram, thus violating our assumption of the convolutional model. Is it *really* bad, however? How bad is it if we take short windows for the estimation of our autocorrelation instead of long ones? If we assume that the wavelet *is* minimum-phase, when it may not be *exactly* minimum-phase, are we really making a bad assumption? Even if every step in the processing of our seismic data adds in noise and is based on very questionable assumptions, does this mean that we cannot see what we are looking for?

Seismic data are known to be robust, and the practicing seismologist, who analyzes and processes hundreds of miles of seismic data every year, knows that most new data-processing techniques, especially in the area of deconvolution, make only marginal improvements to the final stacked section. Very often the so-called improvements that are made are undetectable. The conclusion that is often drawn from all this experience is that there is a limit to the level of sophistication that is required to obtain an adequate seismic section and, further, that basic, normal, everyday seismic processing will yield a product that is not markedly different from what is obtained by much more sophisticated methods. From this observation it is a simple step to the conclusion that basic, normal, everyday seismic processing is fundamentally sound, and that it is pointless to question things that we now know, by experience, to be correct.

In this final chapter I show a seismic section that has been processed in several ways. The differences lie in the way the deconvolution has been done. Overall the sections look very similar, illustrating that the data are robust and look the same whatever is done to them. The main structural features can be seen whatever kind of deconvolution is done.

In the past, most of the oil and gas that has been found has been in structural traps. Therefore, determination of structure has been the most important aim of seismic reflection. Most of the obvious structural traps in the explored areas have now been found; and companies are now looking for less obvious *stratigraphic* traps, which can be detected with the seismic reflection method if local changes in lithology can be detected. These detailed changes in lithology are expressed in the seismic response as lateral changes in amplitude and phase of the earth impulse response $g(t)$. If small changes in $g(t)$ are to be observed from trace to trace in the seismic section, it is essential to be able to depend on the deconvolution.

To see how much in error we can be with what might be regarded as small errors in the assumptions of our deconvolution method, we refer back to the example of section 4.6. In particular, we look at figure 4.12. The upper trace 4.12(1) is the original synthetic white, random, stationary sequence $g(t)$. The lower trace 4.12(5) is the deconvolved estimate of $g(t)$, which looks as if it has approximately the same low-frequency statistics but is locally often completely different. These local errors amount to errors in the determination of lithology.

In fact, of course, both the traces in figure 4.12 are synthetic and, therefore, artificial. Any conclusions we draw from this example may not apply to real data.

In figure 7.1 we show a seismic line from the North Sea containing some obvious structural features. This particular section, entitled Brute Stack, has had no deconvolution applied. The data were shot using air guns as a sound source, with the air guns arranged in four subarrays, each subarray having the configuration illustrated in figure 2.11. Near-field hydrophones, 1 m from each gun as shown in figure 2.11, were monitored and the signatures they received were recorded on tape for every shot. The intention was to derive the far-field signature of the four subarrays from these measurements using the method of Ziolkowski et al. (1982). The data were obtained by the British National Oil Corporation (BNOC, now Britoil Limited), and the sections I show here were processed by Roger Hurley of BNOC.

The basic data were processed in four different ways:

1. Without deconvolution (brute stack).
2. With one-step-ahead predictive-deconvolution assuming the wavelet to be minimum-phase.
3. Using a state-of-the-art wavelet-extraction method.

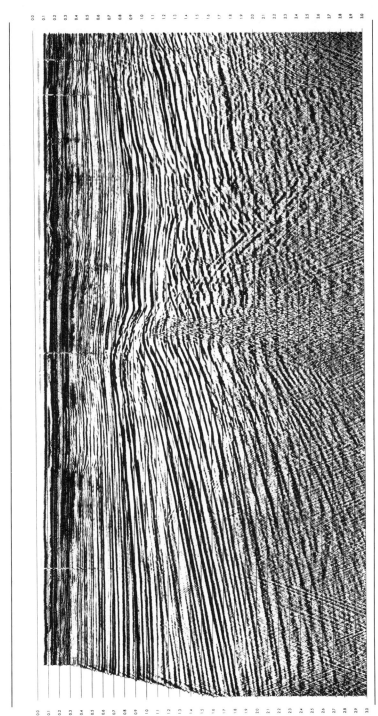

Figure 7.1. A North Sea line–brute stack. Reprinted by permission of Britoil PLC.

131

4. Using a far-field signature determined from near-field measurements by the method of Ziolkowski et al. (1982).

The one-step-ahead predictive-deconvolution method is based on the estimated autocorrelation of the data, trace by trace, assuming that the earth impulse response is white, random, and stationary, and that the wavelet is minimum-phase as described in chapters 3, 4, and 5. Estimation of the spiking deconvolution filter was made on a trace-by-trace basis.

The state-of-the-art wavelet-extraction method is the exponential tapering method described in section 6.3. The best wavelet extracted was then used to derive a filter that would shape the wavelet into a chosen zero-phase wavelet, as described in section 5.4, and this filter was applied to the data. The filter was derived and applied on a shot record–by–shot record basis.

The signatures derived by the method of Ziolkowski et al. (1982) are shown in figure 2.12. They are angular-dependent because the subarrays were directive. In normal data processing this directivity cannot be taken into account, and one signature must be chosen. We chose the vertical incidence (0°) signature shown in figure 2.12. A single filter was designed to shape this chosen wavelet into the same zero-phase wavelet used in the state-of-the-art statistical method. This same filter was applied to every trace in every record along the whole line.

The results of these three different deconvolutions are shown on figures 7.2, 7.3, and 7.4. All other processing parameters were identical. At this scale it is possible to see that the data are robust and that the structural features are not lost whatever deconvolution method is used.

In order to see the effects of the deconvolution on the appearance of the lithology, a portion of the line has been plotted on a larger scale. The four sections shown are: (1) the brute stack in figure 7.5, (2) the spiking deconvolution in figure 7.6, (3) the statistical deconvolution in figure 7.7, and (4) the deterministic signature deconvolution in figure 7.8. A small portion of each of these sections is shown in figures 7.9, 7.10, 7.11, and 7.12, respectively.

The section that best preserves the original character of the brute stack is the deterministic deconvolution based on the near-field measurements. The spiking deconvolution is noisy and breaks up the continuity of the events. The exponential tapering state-of-the-art deconvolution method destroys the basic character of the data and breaks up the continuity of the events.

There is the problem, of course, that we are comparing these sections with each other. We should like to compare them all with "the truth," but in this experiment we do not have the truth. The reader is free to exercise his or her own judgment as to which method yields the most reliable section.

Perhaps the nearest we are ever likely to get to "the truth" is from well information. As mentioned in chapter 2, there are many difficulties in constructing accurate synthetic seismograms from well logs. One aspect of well

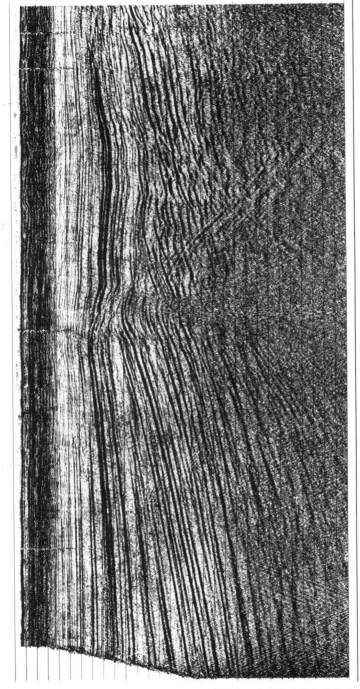

Figure 7.2. A North Sea line-spiking deconvolution stack. Reprinted by permission of Britoil PLC.

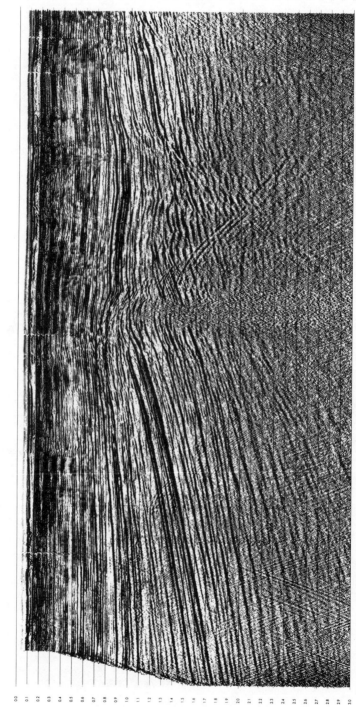

Figure 7.3. A North Sea line–statistical deconvolution stack. Reprinted by permission of Britoil PLC.

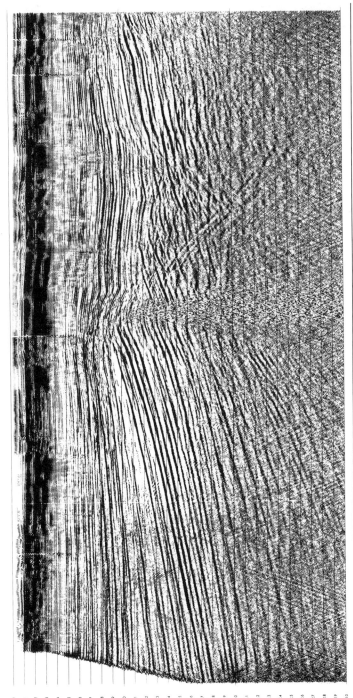

Figure 7.4. A North Sea line—deterministic signature deconvolution stack. Reprinted by permission of Britoil PLC.

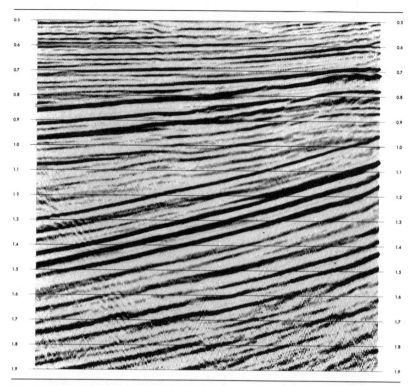

Figure 7.5. A North Sea line–brute stack. Reprinted by permission of Britoil PLC.

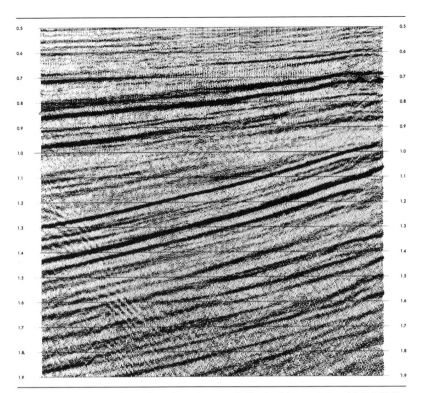

Figure 7.6. A North Sea line–spiking deconvolution stack. Reprinted by permission of Britoil PLC.

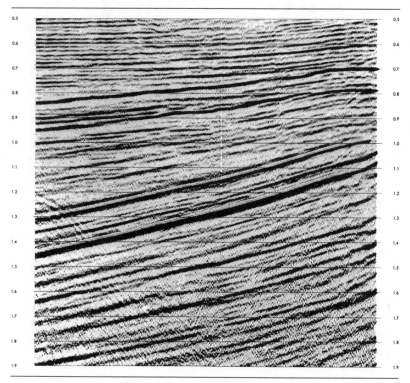

Figure 7.7. A North Sea line–statistical deconvolution stack. Reprinted by permission of Britoil PLC.

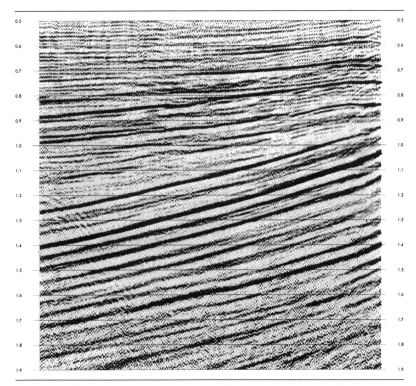

Figure 7.8. A North Sea line–deterministic signature deconvolution stack. Reprinted by permission of Britoil PLC.

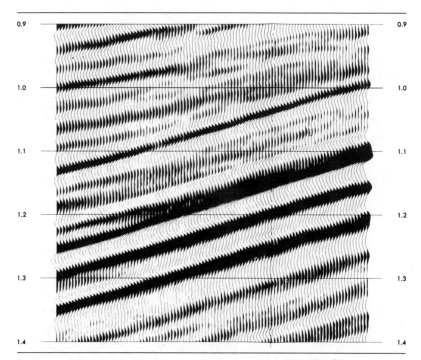

Figure 7.9. A North Sea line–brute stack. Reprinted by permission of Britoil PLC.

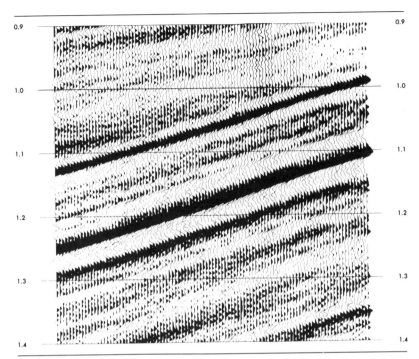

Figure 7.10. A North Sea line–spike deconvolution stack. Reprinted by permission of Britoil PLC.

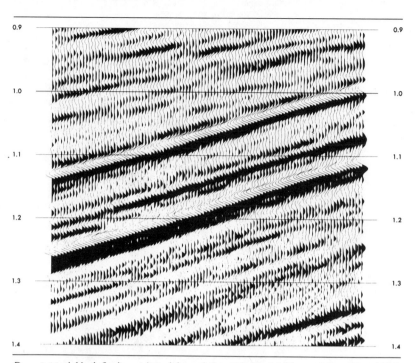

Figure 7.11. A North Sea line—statistical deconvolution stack. Reprinted by permission of Britoil PLC.

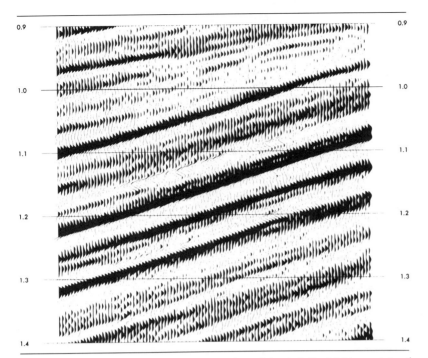

Figure 7.12. A North Sea line–deterministic signature deconvolution stack. Reprinted by permission of Britoil PLC.

logs that is often noticed, however, is that the plane-wave reflection impulse response is not white. In other words $\phi_{gg}(\tau)$ is not an impulse. One way to improve the estimate of the autocorrelation of the wavelet from the data is to divide out the known nonwhite spectrum of the impulse response of the earth. In most places we do not know the impulse response of the earth or its spectrum; this is what deconvolution is supposed to help us to find.

As a final word it should be said that, since the angular dependence of the wavefield can now be calculated from near-field measurements, and it is now known that the wavefield is directional, the point-source convolutional model is inadequate. We must develop methods to cope with this directional wavefield. We must also ensure that the field data are recorded in such a way that the source wavefield may be determined from measurements. That is, we must try to measure what we are doing.

In principle we put a sound wave into the earth and record the echoes. We are trying to find out about the earth by seeing what it does to our input wave. Statistical deconvolution methods attempt to find how the earth alters the input wave without knowing what the input wave is. Knowledge of the input wave is crucial to the solution of the problem.

APPENDIX

A.1. The Impulse or Dirac Delta Function δ(t)

As described by Bracewell (1965), the impulse $\delta(t)$ is more of a concept than a physical reality. It is an infinitely strong pulse of unit area which can be described mathematically as

$$\delta(t) = 0, \quad t \neq 0$$

and $\displaystyle \int_{-\infty}^{\infty} \delta(t)dt = 1.$ \hfill (A.1)

Physically, such a thing does not exist. However, if the impulse response of the system under test does not vary too rapidly with time, it may be possible to create a physical pulse which is brief enough and concentrated enough to behave exactly like the infinitely brief, infinitely concentrated pulse of equation (A.1).

Consider the rectangular function $\tau^{-1}\Pi(t/\tau)$ illustrated in figure A.1. This is a pulse with a width of τ and a height of τ^{-1}. Therefore its area is unity, whatever the value of τ. This rectangular pulse behaves like our ideal pulse as its width tends to zero:

$$\lim_{\tau \to 0} \tau^{-1}\Pi(t/\tau) = 0, \quad \tau \neq 0$$

$$\lim_{\tau \to 0} \int_{-\infty}^{\infty} \tau^{-1}\Pi(t/\tau)dt = 1. \hfill (A.2)$$

So we consider this rectangular pulse as having the properties we desire provided τ is small enough. Whenever we consider the impulse function $\delta(t)$ we can think of this rectangular function with a very small τ.

Now consider the integral:

$$\int_{-\infty}^{\infty} \delta(t)g(t)dt,$$

where $g(t)$ is a function of time. Figure A.2 illustrates $g(t)$ and the integrand is shown by the broken line. Its area is τ^{-1} times the shaded area. The shaded

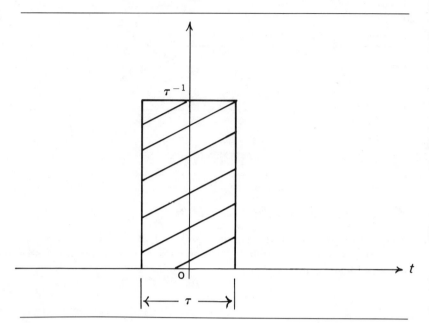

Figure A.1. The rectangular pulse $\tau^{-1}\Pi(t/\tau)$.

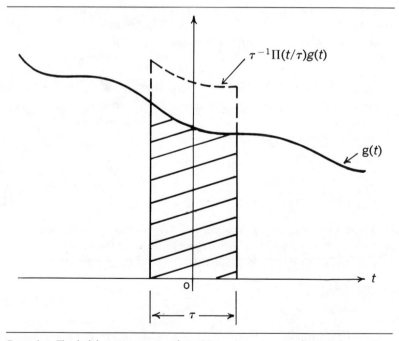

Figure A.2. The shaded area is approximately $\tau g(0)$.

area has a width of τ and an average height of $g(0)$. Thus the area under the integrand approaches $g(0)$ as τ tends to zero and we can deduce that:

$$\int_{-\infty}^{\infty} \delta(t)g(t)dt = g(0). \tag{A.3}$$

Since the characteristics of the pulse shape do not enter the right-hand side of this equation, we see that $\delta(t)$ can be a pulse of any shape; it does not have to be a rectangular pulse. The impulse $\delta(t)$, therefore, simply stands for a unit pulse whose duration is much smaller than any time interval of interest, and we represent it as a spike of height equal to unity, as shown in figure A.3.

Consider now the integral

$$\int_{-\infty}^{\infty} \delta(t-a)g(t)dt.$$

This is illustrated in figure A.4, using the rectangular function, as before. We see that $\delta(t-a)$ is an impulse at time $t = a$ and, therefore, in the case where the width of the rectangular pulse tends to zero, we can see that:

$$\int_{-\infty}^{\infty} \delta(t-a)g(t)dt = g(a). \tag{A.4}$$

A.2. Impulse Response and Convolution

Figure A.5(a) illustrates a physical linear system whose response to the impulse $\delta(t)$ is $h(t)$. We call $h(t)$ the *impulse response* of the system. Physical systems cannot respond before they have an input; that is, they are *causal*. We express this causality of the system mathematically as:

$$h(t) = 0, \quad t < 0. \tag{A.5}$$

If we delay the input of the impulse until time $t = a$, we will also delay the output as illustrated in figure A.5(b). Referred to the same time origin, this delayed impulse response is $h(t - a)$.

Now consider an input $s(t)$ to the same system, as shown in figure A.6, where $x(t)$ is the output. We can calculate $x(t)$ by dividing the input $s(t)$ into little rectangular strips, as shown in figure A.7. The output can be calculated as the sum of the responses to this series of rectangular input pulses. We are entitled to sum the sequence of outputs in this way only if the system is linear. First we define the input strip at time $t = a$, using equation (A.4) on $s(t)$:

$$\int_{-\infty}^{\infty} \delta(t - a)s(t)dt = s(a). \tag{A.6}$$

This is an impulse of area $s(a)$ at time $t = a$. The response to a unit impulse at time $t = a$ is $h(t - a)$, as illustrated in figure A.5(b). Therefore, since the system is linear, the response to our input strip at time $t = a$ is

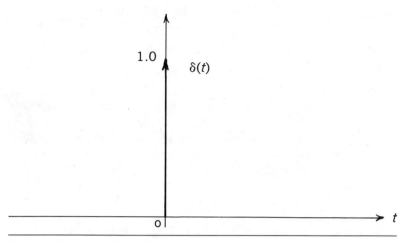

Figure A.3. The impulse $\delta(t)$.

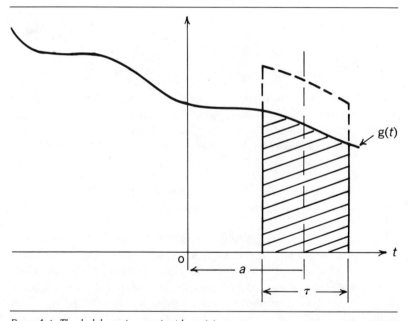

Figure A.4. The shaded area is approximately $\tau\,g(a)$.

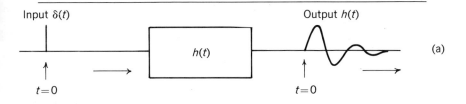

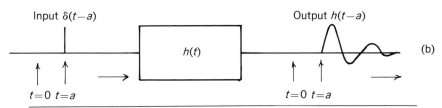

Figure A.5. Impulse Response. (a) Response of system h(t) to impulse at t = 0. (b) Response of system h(t) to impulse at t = a.

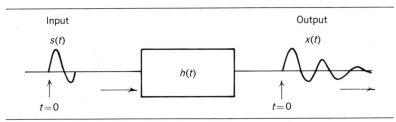

Figure A.6. Response of system h(t) to input s(t).

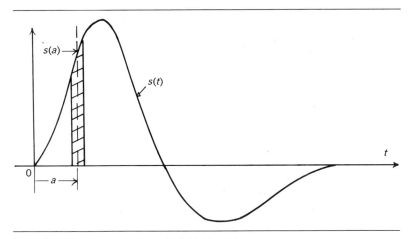

Figure A.7. The input signal s(t) can be divided into strips.

$$s(a)h(t - a).$$

We find the total response $x(t)$ by adding up the responses to all the impulsive strips for all values of a:

$$x(t) = \int_{-\infty}^{\infty} s(a)h(t - a)da. \tag{A.7}$$

Now, since $h(t)$ is causal, it cannot respond before there is an input. Therefore, the output $x(t)$, at time t, is the response to all the inputs up to and including time t; that is, the upper limit of the integral should be t:

$$x(t) = \int_{-\infty}^{t} s(a)h(t - a)da. \tag{A.8}$$

In fact the contribution to the integral in equation (A.7) is zero for values of a greater than t because causality ensures that $h(t - a)$ is zero in this range [see equation (A.5)]. Equations (A.7) and (A.8) are therefore identical for causal systems. By substituting $\tau = t - a$, we can see that the integral in equation (A.8) may be expressed as

$$\begin{aligned} x(t) &= \int_{\infty}^{0} s(t - \tau)h(\tau)(-d\tau) \\ &= \int_{0}^{\infty} s(t - \tau)h(\tau)d\tau \\ &= \int_{-\infty}^{\infty} s(t - \tau)h(\tau)d\tau, \end{aligned} \tag{A.9}$$

where, again, causality ensures that there is no contribution to the integral for τ less than zero.

The integrals in equations (A.7), (A.8), and (A.9) are *convolutions;* and we say that the output $x(t)$ of the linear system is the convolution of the input $s(t)$ with the impulse response of the system $h(t)$. Since equations (A.8) and (A.9) are equal, we notice that the *order* of the convolution is irrelevant. That is, the output response $x(t)$ is the response of the linear system $h(t)$ to the input $s(t)$; but it could equally well be the response of a linear system with impulse response $s(t)$ to an input $h(t)$. The equivalence of these two conceptions is depicted in figure A.8. There is a convenient shorthand for the convolution integral equation which is:

$$s(t) * h(t) = h(t) * s(t) = x(t). \tag{A.10}$$

We note that our earlier descriptions of the impulse response may also be written in this convolutional form; that is

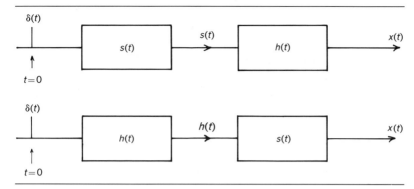

Figure A.8. The order of the convolution is irrelevant.

$$\delta(t) * h(t) = h(t) \tag{A.11}$$

$$\text{and} \quad \delta(t - a) * h(t) = h(t - a). \tag{A.12}$$

As we noted earlier, $\delta(t)$ can be a pulse of any shape provided it is briefer than any time interval of interest. And we further see that, since the impulse response does not depend on the shape of $\delta(t)$, we cannot tell from the impulse response what shape $\delta(t)$ is. We can get a better idea of the constraints on a physical approximation of our ideal impulse by considering the process in the frequency domain.

A.3. The Fourier Transform and the Spectrum of an Impulse

The frequency domain equivalent of the impulse response is known as the *transfer function*. The impulse response $h(t)$ and the corresponding transfer function $H(f)$ are a Fourier transform pair:

$$H(f) = \int_{-\infty}^{\infty} h(t)e^{-2\pi ift}dt \tag{A.13}$$

$$h(t) = \int_{-\infty}^{\infty} H(f)e^{2\pi ift}df. \tag{A.14}$$

The transfer function $H(f)$ has both amplitude and phase, both of which vary with frequency f. Thus we may write:

$$H(f) = |H(f)|e^{i\theta(f)}, \tag{A.15}$$

where $|H(f)|$ is the amplitude spectrum and $\theta(f)$ is the phase spectrum. We show these two components in figure A.9.

We may find the spectrum, or transfer function, of an impulse by applying the result of equation (A.3) to the integral of equation (A.13), substituting $\delta(t)$ for $h(t)$ and $e^{-2\pi ift}$ for $g(t)$ to yield:

$$\int_{-\infty}^{\infty} \delta(t)e^{2\pi ift}dt = 1. \tag{A.16}$$

Thus the impulse $\delta(t)$ has a unit amplitude spectrum and a zero phase spectrum, as shown in figure A.10.

A.4. The Convolution Theorem

The convolution theorem states that the Fourier transform of the convolution of two functions is equal to the multiplication of the Fourier transforms of the two functions. That is, if $H(f)$ is the Fourier transform of $h(t)$ and $S(f)$ is the Fourier transform of $s(t)$, then $s(t) * h(t)$ has the Fourier transform $S(f) \cdot H(f)$. We may write this as:

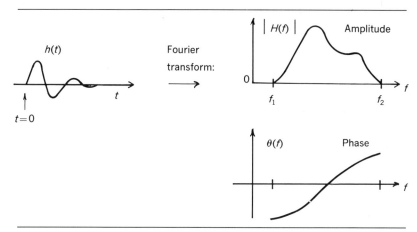

Figure A.9. The Fourier transform of the impulse response h(t) *is the transfer function,* H(f) *which has both amplitude* | H(f) | *and phase* θ(f).

$$FT[s(t) * h(t)] = S(f) \cdot H(f).$$ (A.17)

We can prove equation (A.17) as follows:

$$\int_{-\infty}^{\infty} \left[\int_{-\infty}^{\infty} s(a)h(t-a)da\right]e^{-2\pi ift}dt$$

$$= \int_{-\infty}^{\infty} s(a)\left[\int_{-\infty}^{\infty} h(t-a)e^{-2\pi if(t-a)}dt\right]e^{-2\pi ifa}da$$

$$= \int_{-\infty}^{\infty} s(a)H(f)e^{-2\pi ifa}da$$

$$= H(f)\int_{-\infty}^{\infty} s(a)e^{-2\pi ifa}da$$

$$= H(f) \cdot S(f).$$ (A.18)

The frequency domain representation of the impulse response test of figure A.5(a) is shown in figure A.11 and is obtained by applying the convolution theorem to equation (A.11):

$$FT[\delta(t)] \cdot H(f) = H(f)$$ (A.19)

The transfer function $H(f)$ is a complex function of frequency containing an amplitude part $|H(f)|$ and a phase part $e^{i\theta(f)}$. Multiplication of two Fourier transforms can be treated as an arithmetic multiplication of amplitudes, frequency by frequency, and addition of the phases, frequency by frequency. Since the impulse $\delta(t)$ has an amplitude spectrum which is equal to 1.0 at all frequencies and a phase spectrum which is 0.0 for all frequencies, the complex multiplication of equation (A.19) leaves the transfer function unchanged, and the output is the same as the transfer function.

It is clear from figure A.11 that the impulse applied need not have a flat amplitude spectrum from 0.0 out to infinity. The output of this system is only detectable in the range f_1 to f_2. The frequency components in the impulse below f_1 and above f_2 contribute nothing to our knowledge of the system and it would not matter if they were missing. Indeed it is not possible to tell at the output if they are missing or not. So any pulse which has a flat spectrum in the range f_1 to f_2 will have sufficient *bandwidth* for the system under test.

The amplitude spectrum of the impulse $\delta(t)$ is 1.0 at all frequencies. Such a spectrum is an example of a *white* spectrum. A white spectrum is an amplitude spectrum which is constant at all frequencies.

A.5. Continuous versus Discrete Time

A signal which varies continuously with time may be sampled at regular discrete time intervals Δt. The discrete signal may be used to reconstruct the

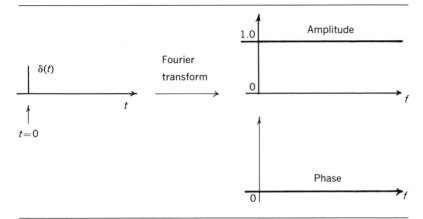

Figure A.10. The Fourier transform of the impulse δ(t) is 1.

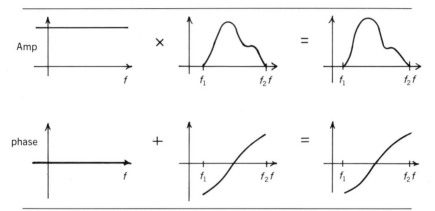

Figure A.11. Frequency domain representation of impulse response test of figure A.5(a).

original continuous signal. If this reconstruction is exact, the sampling is adequate. There are still some mysteries to be solved concerning the correspondence between discrete and continuous signals which we shall not explore here. We simply assume that it is possible to sample continuous signals adequately and that normal digital seismic data consists of signals varying continuously with time which have been sampled at an adequately small interval Δt. The relationship between a continuous signal $x(t)$ and its discrete counterpart x_k is

$$x_k = \int_{-\infty}^{\infty} x(t)\delta(t - k\Delta t)dt; \tag{A.20}$$

that is, the discrete signal x_k exactly matches the continuous signal $x(t)$ at times corresponding to the integer index k.

In discrete time the suffix or index denotes the number of the sample. For example, the discrete impulse, or kroneker delta δ_t, is similar in behavior to the impulse $\delta(t)$ in continuous time. The discrete impulse is defined as:

$$\begin{aligned} \delta_t &= 0, \quad t \neq 0 \\ &= 1, \quad t = 0. \end{aligned} \tag{A.21}$$

Convolution of two signals s_t and h_t is a summation:

$$\begin{aligned} x_t &= \sum_{\tau=-\infty}^{\infty} s_\tau h_{t-\tau} \\ &= \sum_{\tau=-\infty}^{\infty} s_{t-\tau} h_\tau. \end{aligned} \tag{A.22}$$

The Fourier transform for discrete signals becomes the discrete Fourier transform (*DFT*):

$$X(f) = \sum_{k=0}^{N-1} x_k e^{-2\pi i f k \Delta t}. \tag{A.23}$$

The sequence x_k contains N samples which may be complex; therefore $2N$ (real and imaginary) independent pieces of information are required to specify x_k. To ensure that the *DFT* also contains $2N$ independent pieces of information, it is necessary to specify $X(f)$ at N independent frequencies f which are integer multiples of some frequency Δf. Since the complex exponential function is periodic, it is normal to choose Δf such that the independent frequencies are uniformly distributed over one period; that is,

$$\Delta f = \frac{1}{N\Delta t} \tag{A.24}$$

and $f = n\Delta f, \quad n = 0, 1, \ldots, N.$ \hfill (A.25)

We may therefore write the forward *DFT* as

$$X(f) = X_n = \sum_{k=0}^{N-1} x_k e^{-2\pi i n k/N} \tag{A.26}$$

while the inverse *DFT* becomes

$$x_k = \frac{1}{N} \sum_{n=0}^{N-1} X_n e^{2\pi i n k/N}. \tag{A.27}$$

Certain concepts in continuous time do not have obvious counterparts in discrete time. Causality is one such concept. In continuous time the definition of causality is straightforward (see Section A.2). In discrete time, the periodicity of the *DFT* ensures that any finite-length discrete signal is periodic, and therefore not causal. The problem can be appreciated by considering the associated problem of convolution of two discrete finite-length sequences s_t and h_t, where

$$s_t = s_0, s_1, \ldots, s_n \tag{A.28}$$

$$h_t = h_0, h_1, \ldots, h_m. \tag{A.29}$$

The convolution can be written as

$$x_t = \sum_{\tau=0}^{n} s_\tau h_{t-\tau} \tag{A.30}$$

or $x_t = \sum_{\tau=0}^{m} h_\tau s_{t-\tau}.$ \hfill (A.31)

Using either of these two expressions we find that x_t is a sequence of length $m+n+1$:

$$x_t = x_0, x_1, \ldots, x_{m+n}. \tag{A.32}$$

This convolution could be accomplished by multiplication in the frequency domain using the convolution theorem:

$$DFT(s_t) \cdot DFT(h_t) = DFT(x_t). \tag{A.33}$$

However, a little care is required, because s_t is a $n+1$-length signal, h_t is a $m+1$-length signal and x_t is a $n+m+1$-length signal. The lowest frequency of interest in this equation is

$$\Delta f = \frac{1}{n+m+1}; \tag{A.34}$$

all the other frequencies in the *DFT*s must be integer multiples of this frequency. To ensure that this is so, we must add m zeros to the signal s_t to make it of length $n+m+1$, and n zeros to h_t, to make it of the same length. With the addition of these zeros the multiplication in the frequency domain may be performed correctly, yielding a result which is indeed the *DFT* of the signal which is obtained by convolution in the discrete time domain.

A.6. Transient Signals, Infinitely Long Signals and Stationarity

A *transient signal* is one which exists for a finite length of time. It has a beginning and an end, and may be represented by a finite number of regular discrete samples. For example, if $s(t)$ is a transient signal, it may be represented by the discrete $n+1$-length sequence

$$s_t = s_0, s_1, \ldots, s_n. \tag{A.35}$$

An infinitely long signal exists for an infinite length of time. Such signals may be represented by an infinite number of regular discrete samples. For example, the signal may have a beginning, but no end:

$$x_t = x_0, x_1, x_2. \ldots \tag{A.36}$$

Or it may have an end, but no beginning:

$$y_t = \ldots, y_{-2}, y_{-1}, y_0. \tag{A.37}$$

Or it may have no beginning and no end:

$$z_t = \ldots, z_{-1}, z_0, z_1, \ldots. \tag{A.38}$$

An important class of this third kind of infinitely long sequence is the stationary time series. A *stationary time series* is one whose statistical properties do not change with time. Signals generated by a stationary process have the property that a long enough segment of any signal recorded during some past time period has the same statistics as another segment of the signal observed at a later time. That is, the statistical properties of the signal do not depend on the time origin of the observation. If none of the probabilities which characterize the process changes with time, the process is said to be *stationary in the strict sense*. There is a less strict definition of stationarity known as stationarity in the wide sense. If the mean value of the signal at different times is constant, and if the mean of the product of the values of the signal at two instants of time does not depend upon the absolute time, but only on the time difference, then the process is said to be *stationary in the wide sense*.

In geophysics it is often assumed that seismic reflection signals possess the property of wide-sense stationarity.

Transient signals have *energy* equal to the sum of the squares of the discrete coefficients. Thus the transient signal s_t of equation (A.35) has energy

$$E_s = \sum_{t=0}^{n} s_t^2, \tag{A.39}$$

and since n is finite and all the coefficients are finite, the energy of a transient signal must be finite.

It follows that the energy of an infinitely long signal is infinite, since it is the sum of an infinite number of finite positive terms. However, the *power* of an infinitely long signal is finite. The power is defined as:

$$P_x = \lim_{T \to \infty} \frac{1}{2T+1} \sum_{t=-T}^{T} x_t^2. \tag{A.40}$$

The *mean value* or *mean* of an infinitely long sequence x_t is

$$\bar{x} = \lim_{T \to \infty} \frac{1}{2T+1} \sum_{t=-T}^{T} x_t. \tag{A.41}$$

Two other important statistical properties of infinitely long sequences are the standard deviation σ and the variance σ^2, which is simply the square of the standard deviation. Both are defined by

$$\sigma^2 = \lim_{T \to \infty} \frac{1}{2T+1} \sum_{t=-T}^{T} (x_t - \bar{x})^2. \tag{A.42}$$

We note that the variance σ^2 and the power P_x are the same when the sequence has zero mean.

A.7. Autocorrelation and Cross Correlation

The autocorrelation of a complex sequence x_t is

$$\phi_{xx}(\tau) = \sum_{t=\tau}^{\infty} x_t x^*_{t-\tau}, \tag{A.43}$$

where the asterisk denotes complex conjugate and we follow Robinson's (1967) notation of putting the shift τ in parentheses. Figure (A.12) illustrates the complex autocorrelation function, and shows also that that it exhibits complex conjugate symmetry. We may see this mathematically by substituting $k = t - \tau$ in equation (A.43):

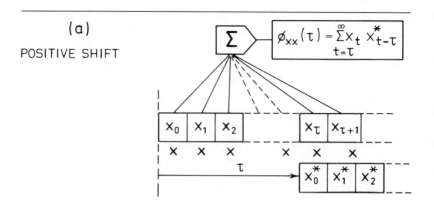

POSITIVE SHIFT

$$\phi_{xx}(\tau) = \sum_{t=\tau}^{\infty} x_t \, x_{t-\tau}^*$$

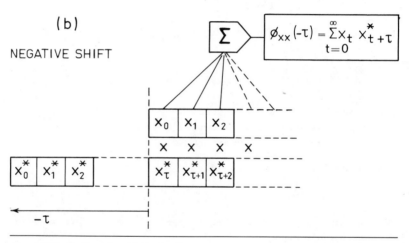

(b)

NEGATIVE SHIFT

$$\phi_{xx}(-\tau) = \sum_{t=0}^{\infty} x_t \, x_{t+\tau}^*$$

Figure A.12. Complex autocorrelation function.

$$\phi_{xx}(\tau) = \sum_{k=0}^{\infty} x_{k+\tau} x_k^*$$

$$= \phi_{xx}^*(-\tau). \tag{A.44}$$

If x_t is real, the autocorrelation is real and perfectly symmetrical; that is

$$\phi_{xx}(\tau) = \phi_{xx}(-\tau). \tag{A.45}$$

If the sequence x_t is infinitely long, the autocorrelation should be redefined as:

$$\phi_{xx}(\tau) = \lim_{T \to \infty} \frac{1}{2T+1-\tau} \sum_{t=-T+\tau}^{T} x_t x^*_{t-\tau} \tag{A.46}$$

The cross correlation of two complex sequences x_t and y_t is defined as

$$\phi_{xy}(\tau) = \sum_{t=\tau}^{\infty} x_t y^*_{t-\tau}. \tag{A.47}$$

Or as

$$\phi_{xy}(\tau) = \lim_{T \to \infty} \frac{1}{2T+1-\tau} \sum_{t=-T+\tau}^{T} x_t y^*_{t-\tau}. \tag{A.48}$$

The cross correlation function does not exhibit any kind of symmetry. That is, it matters whether the shift τ is positive or negative.

A.8. White, Random, and Stationary Time Series

A *white* sequence is one whose amplitude spectrum is a constant for all frequencies, as described in section A.3. In the case of discrete signals this must be taken to mean that the amplitude of every complex value of the discrete Fourier transform (*DFT*) is the same. We note that the *DFT* can be calculated only for a finite number of samples (see section A.5).

A *stationary* time series is infinitely long and has statistical properties which are independent of time (see section A.6). If x_t is stationary in the wide sense, it has a constant mean $\bar{x}$ [see equation (A.41)] and a constant autocorrelation function $\phi_{xx}(\tau)$ [see equation (A.45)].

If a series is *white and stationary* it must have all the above properties. We can see how they are related by considering the spectrum of x_t:

$$X(f) = |X(f)| e^{i\theta_x(f)}, \tag{A.48}$$

where $|X(f)|$ is the amplitude spectrum and $\theta_x(f)$ is the phase spectrum. If we now multiply $X(f)$ by its complex conjugate we form the *power spectrum*.

$$\Phi_{xx}(f) = X(f) \cdot X^*(f) = |X(f)|e^{i\theta_x(f)} \cdot |X(f)|e^{-i\theta_x(f)}$$
$$= |X(f)|^2 \tag{A.50}$$

which is real, and has no phase. If $|X(f)|$ is a constant, $|X(f)|^2$ is a constant, the power spectrum $\Phi_{xx}(f)$ is a constant and is indistinguishable from the Fourier transform of an impulse $\delta(t)$, apart from a scaling factor. In fact the power spectrum $\Phi_{xx}(f)$ is the Fourier transform of the autocorrelation $\phi_{xx}(\tau)$. [This will not be proved for infinitely long sequences, but it is a consequence of the Wiener-Khintchine theorem; see Robinson (1967).] It follows that the autocorrelation of a white stationary sequence is a scaled impulse function $\delta(\tau)$:

$$\phi_{xx}(\tau), = P_x\delta(\tau), \tag{A.51}$$

where P_x is the average power of the sequence x_t and is equal to the zero-lag autocorrelation coefficient. If x_t has zero mean, then P_x may be replaced by the variance σ^2 (see section A.6). We note that the behavior of the phase spectrum is not specified by whiteness or stationarity.

If a sequence x_t is white, *random,* and stationary, the randomness can only describe the phase spectrum $\theta_x(f)$. There is sometimes a confusion in the geophysical literature between whiteness and randomness and it is worth a few words to explain the difference.

Suppose we have a box containing n identical pieces of paper. Each piece of paper has a number written on it. If we shake the box, pick out a piece of paper at random, read the number and then put the paper back, we do not change the distribution of numbers in the box. If we repeat this process over and over again, the resulting sequence of numbers is a random number sequence. The actual numbers in the sequence are the numbers that are in the box. The autocorrelation of the sequence depends on distribution of numbers in the box. It could be, for example, that two pieces of paper have the same number written on them, while all the other numbers might be different. Clearly the *distribution* of numbers is a completely separate issue from the *process* which is used to select the numbers from the distribution.

The phase of a signal at any frequency can lie in the range $-\pi$ to π. The distribution of phases in this range might be uniform, exponential, Gaussian, Poisson, or anything else. A *white, random, stationary* time series is a stationary time series with a white power spectrum and a random phase spectrum; the *distribution* of phases is unspecified.

A.9. The z-Transform and Convolution

Time series are often represented by their z-transforms. The z-transform of a sequence x_t is defined as

$$X(z) = \sum_{t=-\infty}^{\infty} x_t z^t,$$ (A.52)

where multiplication by z corresponds to a time delay of one sample interval, multiplication by z^2 corresponds to a time delay of two sample intervals and so on; similarly, multiplication by z^{-1} corresponds to a time advance of one sample interval, multiplication by z^{-2} corresponds to a time advance of two sample intervals, and so on.

Multiplication of z-transforms corresponds to convolution in the time domain. We may see this by considering the convolution of the two finite-length sequences

$$a_t = a_0, a_1, \ldots, a_n$$

and $b_t = b_0, b_1, \ldots, b_m.$

Their convolution is

$$c_t = \sum_{\tau=0}^{n} a_\tau b_{t-\tau} = \sum_{\tau=0}^{m} b_\tau a_{t-\tau}$$ (A.53)

$$= c_0, c_1, c_2, \ldots, c_{n+m},$$

in which we see that

$$c_0 = a_0 b_0,$$

$$c_1 = a_1 b_0 + a_0 b_1,$$

and so on. The z-transforms of a_t and b_t are

$$A(z) = a_0 + a_1 z + \ldots + a_n z^n$$

$$B(z) = b_0 + b_1 z + \ldots + b_m z^m.$$

The product of these two polynomials is

$$A(z)B(z) = a_0 b_0 + (a_1 b_0 + a_0 b_1)z + \ldots + a_n b_m z^{n+m},$$ (A.54)

with the general term being of the form

$$\sum_{\tau=0}^{n} a_\tau b_{t-\tau} z^t.$$

We see that the coefficients of the z-transform on the right-hand side of equation (A.54) are the same as the samples in the time series c_t, and therefore

$$A(z)B(z) = C(z).$$ (A.55)

This result is very similar to the convolution theorem, suggesting that the z-transform and Fourier transform are closely related, which they are of course. This relationship is discussed in chapter 3.

REFERENCES

Aki, K. and Richards, P.G., 1980, Quantitative Seismology, Theory and Methods, Vol I, W. H. Freeman and Co., San Francisco.

Berkhout, A.J., 1970, Minimum phase in sampled signal theory: Ph.D. thesis, Koninklijke Shell Exploration and Production Laboratory, Rijswijk, Netherlands.

———1974, Related properties of minimum-phase and zero-phase time functions: Geophysical Prospecting, v. 22, p. 683–709.

Berkhout, A. J., and Zaanen, P. R., 1976, A comparison between Wiener filtering, Kalman filtering and deterministic least squares estimation: Geophysical Prospecting, v. 24, p. 141–197.

Bracewell, R., 1965, The Fourier transform and its applications: New York, McGraw-Hill Co.

Burg, J. P., 1967, Maximum entropy spectral analysis: Presented at the 37th Annual International SEG Meeting, in Oklahoma City.

———1972, The relationship between maximum entropy spectra and maximum likelihood spectra: Geophysics, v. 37, p. 375–376.

———1975, Maximum entropy spectral analysis: Ph.D. thesis, Department of Geophysics, Stanford University.

Choy, G. L., and Richards, P. G., 1975, Pulse distortion and Hilbert transformation in multiply reflected and refracted body waves: Bulletin of the Seismological Society of America, v. 65, p. 55–70.

Claerbout, J. F., 1976, Fundamentals of geophysical data processing: New York, McGraw-Hill Co.

Dunkin, J. W., and Levin, F. K., 1973, Effect of normal movement on a seismic pulse: Geophysics, v. 38, no. 4.

Goupilland, P.L., 1961, An approach to inverse filtering of near-surface layer effects from seismic records: Geophysics, v 26, p. 754–760.

Hughes, V. J. and Kennett, B.L.N., 1983, The nature of seismic reflections from coal seams, First Break, vol 1, No. 2, pp. 9–18.

Levinson, N., 1947, The Wiener R.M.S. (root mean square) error criterion in filter design and prediction: Appendix B in Weiner, N., Time series, 1964: Cambridge, Mass., MIT Press.

Mayne, W. H., 1962, Common reflection point horizontal data stacking techniques: Geophysics, v. 27, no. 6.

Peacock, K. L., and Treitel, S., 1969, Predictive deconvolution: theory and practice: Geophysics, v. 34, no. 2.

Popper, K. R., 1972, Conjectures and refutations: London, Routledge and Kegan Paul.

Robinson, E. A., 1957, Predictive decomposition of seismic traces: Geophysics, v. 22, p. 767–778.

——— 1962, Random wavelets and cybernetic systems: London, Charles Griffin and Co.

———1967, Statistical communication and detection: London, Charles Griffin and Co.

Robinson, E. A., and Treitel, S., 1967, Principles of digital Wiener filtering: Geophysical Prospecting, v. 15, p. 311–333.

———1980, Geophysical signal analysis: Englewood Cliffs, N.J., Prentice-Hall.

Stoffa, P. and Ziolkowski, A., 1983, Seismic source decomposition: Geophysics, v.48, p.1.

Treitel, S., and Robinson, E. A., 1966, Seismic wave propagation in terms of communication theory: Geophysics, v. 31, p. 17–32.

Webster, G. M., 1978, Deconvolution, vols. 1 and 2; Geophysics reprint series, SEG, P.O. Box 3098, Tulsa, Okla. 74101.

Wiener, N., 1964, Time series: Cambridge, Mass., MIT Press.

Ziolkowski, A., 1982, Further thoughts on Popperian geophysics—the example of deconvolution: Geophysical Prospecting, v. 30, p. 155–165.

Ziolkowski, A., and Lerwill, W. E., 1979, A simple approach to high resolution seismic profiling for coal: Geophysical Prospecting, v. 27, p. 360–393.

Ziolkowski, A., Parkes, G., Hatton, L., and Haugland, T., 1982, The signature of an air gun array: Computation from near field measurements including interactions: Geophysics, v. 47, p. 1413–1421.

Index